THE KEEP

The Keep

Living with the Tame and the Wild on a Mountainside Farm

Henry T. Ireys and
Priscilla M. Ireys

WEST VIRGINIA UNIVERSITY PRESS · MORGANTOWN

Library of Congress Control Number: 2025004010

"Izzy's Bridge" was previously published in the *Anthology of Appalachian Writers, Marc Harshman Volume 17* (Shepard University, May 2025).
A slightly modified version of "Morels on the Mountain" was published in the *Anthology of Appalachian Writers, Barbara Kingsolver Volume XV* (Shepard University, 2023).
The following stories appeared in the *Small Farmer's Journal*:
"Hercules" in volume 46, number 3, 183rd edition
"The Hat of Shame" in volume 46, number 2, 183rd edition
"Pedro" in volume 46, number 1, 181st edition
"Arnost and the Eagle" in volume 45, number 3, 179th edition
"Tiny Tim" in volume 44, number 4, 176th edition

Joe Razes kindly allowed us to use his photographs, which appear on pages 14, 46, 80, 144, and 267.
Lisa Elmaleh kindly allowed us to use her photographs, which appear on pages 73, 186, and 278.

For EU safety/GPSR concerns, please direct inquiries to WVUPress@mail.wvu.edu or our physical mailing address at West Virginia University Press / PO Box 6295 / West Virginia University / Morgantown, WV, 26508, USA.

Cover design by Tatiana Zakharova / WVU Press
Book design by Than Saffel / WVU Press
Cover image courtesy of Joe Razes

For Grandma, Vanessa, Steve, Izzy, Charlie, Nick, Perri, Alex, Su, Roy, Claire, and Ethan.
See you at the farm!

Keep (noun).
The strongest or central tower of a castle, acting as a final refuge.

Contents

....

"Say What? No Way!"

Hard Times

Gifts

Quiet Times

Epilogues

Preface

....

This book evolved from two decades of living on eighty acres in central Appalachia, including ten years of farming with heritage goats, cattle, and hogs. Back then, we would relate a day's events to friends and visitors, and our stories would trigger their memories, and they would tell their tales, and the conversation would meander on, sometimes ending in laughter and sometimes in tears. Stories of eagle attacks, cockeyed goats, intelligent hogs, stunning sunsets, and dreadful mistakes expanded our community in unexpected ways.

Eventually, we wrote our stories down to preserve the sharp edges of our experiences and not let time blunt the memories of despair, hope, humor, surprise, fatigue, sorrow, and joy that came along with running a small farm surrounded by deep forest. At some point, the idea of assembling these pieces between two covers took root and grew like a trumpet vine flourishing haphazardly along a fence: after a while, you can't ignore its boisterous yellow horns.

The trouble was that we often had quite different takes on the same event. In a deer's death during a hunt, one of us saw the demise of a beautiful animal; the other, the gift of meat, gratefully received. When things went wrong, we often found the reason in different places. A few discussions—some might say arguments— led us to realize our life on the farm could not be described in a single linear arc. Writing this book together reminded us there is never one true story.

As we wrote and read our stories to each other, the voice of the land itself became more distinct. It told us things about itself: that it contained more than we could imagine within its boundaries, that it had seen many human comedies long before ours, that it asked us to care for it as best as we could manage. Listening to the land—and telling stories about it—provided pathways for discovering a little bit more about ourselves and our marriage.

We hope this book encourages readers to tell their own tales about journeys in and around special places. In doing so, perhaps they'll discover a few unexpected corners in their own characters.

Introduction

·······

Long before the goats arrived, before we built the barn, before we dreamed of raising livestock, Priscilla and I walked across the land's dormant pastures and along its forested creeks. Before we worried about drought or parasites or coyotes, the land sang to us. We heard its soft, wind-borne tunes—insistent, compelling melodies. With every step under forested canopies and through swaying grasses, its songs invited us to climb and look, to smell, and taste, and listen.

The place asked us to join the chorus, to sing out and soothe our restless souls. It promised to become our "keep," a strong, secure place that would protect its livestock—and its humans. From the beginning, it has kept us as much as we have kept it.

For several years before we found the farm, Priscilla and I had been searching for a place that could meet an insistent need for solitude, for time with just our family, and for some freedom from demanding careers, from the internet's sharp claw, and from a constantly humming city. My mother died late in 2000, and several months later, we found ourselves with a small inheritance. We talked for a long time about how to use it in a way that would make sense to her.

Land would make sense, but not just any land. What was important? Priscilla and I talked back and forth, each insisting that the land had to have certain characteristics: forests, fields, views, gardens, garages, and so on—a ludicrously long catalog of wishes. Finally, we whittled the list down to three essential features: the place had to be off a road that had no line painted down its middle; it had to be up a dirt road with no other houses in sight; and it had to have a house with a working indoor bathroom.

Serendipity struck on a Saturday morning in July 2001 as we returned to our home in Baltimore, Maryland, from our yearly vacation. For a decade, we had rented a log cabin at Watoga State Park in southern West Virginia—a ritual that allowed our two school-aged sons to find crawfish in the nearby creeks, learn to swim in a pool fed by a cold mountain stream, and be at home in the natural world. These vacations reconnected Priscilla and me to important moments in our past and renewed our hope that beautiful places would survive a splintering world.

On that July morning, we found ourselves sitting in a motionless car, stuck on the interstate in a major traffic jam. Ignoring our sons arguing in the backseat, Priscilla picked up a thin local newspaper she'd bought earlier at a tiny grocery store outside a small town in Hampshire County, West Virginia. She opened the paper to the classified section and immediately found a listing for a property near that grocery store: eighty-plus acres of forest and pasture with an old farmhouse. She called the phone number. A man answered, asked us a few questions, and said he'd be at his place the next day. We agreed to meet him there.

The traffic finally started flowing and we arrived home late in the evening. The next day, we left our children with a neighbor and drove to the property. The owner's directions were based on mileage from specific landmarks because this was before smart phones with GPS. The address was just the number on the mailbox at the bottom of the driveway. Before we arrived at that mailbox, we meandered through a dense forest of pines and hardwoods along a narrow, twisting road with no line painted down its middle.

Turning off the paved road at the designated mailbox, we drove up a rutted dirt lane for about half a mile through the forest. We came to an old farmhouse at the edge of a large pasture. As we slowed down to stop in front of its porch, the road's dust slowly settled behind us. We climbed out of the car, stood looking around for a moment, and could see no other homes. Just the pasture and the forest.

The walk through the farmhouse took only a few minutes because it was barely more than a tiny 1930s cabin built on locust posts. Over its lifetime, it had settled into a comfortable lopsidedness. The bathroom approximated a closet, but everything worked. That covered the last item on our list.

And then we walked the land. We walked for several hours, leading with our senses. We heard its varied voices: the wind through the tall grasses, the rustling of thick leaves above us, the soft, barely audible phfft of falling pine needles. We walked through invisible waves of perfumes: the trailing scents of cedar, moldering logs, and autumn olive. In the dark woods surrounding the pastures, we stepped across jagged rocks lying in shallow streams under thick canopies of sixty-foot-tall hardwoods. One of the back pastures held a small, spring-fed pond that promised water for livestock during droughts. On the afternoon of our walk, the surface of its perfectly still water reflected a single cloud in a blue sky.

As we returned to the farmhouse from our walk, I looked back up the hill toward the mountain behind the land: thick woods empty of human houses but home to creatures who live there still. Deer, turkeys, coyotes, fishers, mountain lions, bears, and eagles.

That first walk on the farm reminded me of a 1925 photograph of eleven members of my family—grandparents and great uncles and aunts and my mother, fourteen years old in the picture. They stand stiffly in wide trousers and slim skirts near a post-and-rail fence on New England land that, even then, had already been in the family for three generations. My grandparents had tended milk cows there, long before I was conceived in a back bedroom of the old farmhouse. Most of the people in that picture were dead when I was born in 1952, but that picture connected me to an old place that had been a well-worn and deeply loved stage on which my family had pursued their hopes and dodged their fears for more than a century.

Out of practical necessity, and burdened with much sadness, my sisters and I sold that house and land after our mother's death. But during my first walk around the West Virginia farm, I saw that I had a chance, unbidden and unexpected, to love a place that was textured and resonant enough to gather stories for a new century and for a different version of the same family. I usually don't do things without thinking about them for a long time—too long, sometimes—but on that walk, the land spoke directly to my body, coupling past and future.

The land connected Priscilla to her past, too. Most of her school time was in Georgia, but she spent summers with her father's mother, known in the family as Mamaw, on her farm in Eastern Kentucky—Bear Hollow, to be exact. Priscilla's upstairs bedroom window in Mamaw's house faced a wide pasture and one of her favorite morning treats was to sit in bed and watch the mist rise. The wispy, ghostlike cloud would cover the mountains behind the barn and roll down the hollow. The first morning after moving full-time to our farm, Priscilla stood behind our house and watched the mist rise off the fields. It covered the mountains before disappearing into sunlight. On misty mornings, Priscilla goes back to Mamaw's house. Sometimes, Mamaw's spirit visits our farm.

In the end, we bought the place because it met the demands of our restless imaginations. It was a place to calm our souls and ride our horses. Our sons could build treehouses, sleep under the stars, and learn the power of dreams. This would become our keep.

For a few years after we bought the land, we visited it only during the summer and holiday weekends because other parts of our lives—careers, family, friends—claimed our attention. But in the summer of 2006, we came to stay.

Five years later, we had a core herd of goats: fifty heritage does (mostly Spanish and a few Savanna) for meat production and a stable of five or six bucks—plus three guardian dogs for the herd, another two to protect the house, and a pair of horses for long-distance riding. We bred our goats in the fall, tended to spring kidding, and encouraged late-summer weaning. Later, we tended a small herd of cattle and watched four-month-old piglets grow into mature sows.

We sold our goats to families who needed goat meat for religious festivals, to homesteaders who were raising their own protein, and to other breeders who needed to expand their herds' genetic diversity with heritage lines. Breeders heard about the quality of our goats or read our website and came to purchase them from farms in neighboring states and as far away as Florida, Indiana, and Rhode Island.

Recognizing the farm's breeding success with rare heritage goats, in 2012, the Swiss Village Farm (SVF) Foundation, a non-profit organization that sought to preserve rare breeds of live-stock, asked Priscilla to sell the organization a buck and a group of does from our Spanish herd. The Foundation's staff wanted to gather semen and embryos and store them in an animal "seed bank." She was delighted to have her goats selected as strong representatives of the Spanish goat breed. In 2013, Animal Welfare Approved (AWA), an independent accrediting orga-nization designed to rate the degree to which farms treat their animals humanely, gave our farm its highest ranking—a rare achievement.

While Priscilla tended to the farm, I kept my job at a con-sulting firm in Washington, DC, managing large research and evaluation projects within the nation's health care system. For about five years after we moved full time to the farm, I spent three days a week working in DC and two from the farm's office. When I was home on weekday evenings and weekends, I'd join Priscilla in the barns and fields or manage the administrative side of the farm's operations.

Like all farms, ours craved attention. Goats became ill suddenly; fences fell over; a barn's roof blew off; a storm unex-pectedly dropped enough snow to block pasture gates; pipes and water troughs froze; electricity failed. To Priscilla, major crises seemed to happen when I wasn't around, and she often felt my absence acutely. To me, major crises seemed to happen all the time. Driving home from my office, I'd wonder what new problems were waiting and how I'd manage to finish my own work and the farm's.

As its reputation grew, the farm laid the foundation for its legacy: Priscilla's breeding lines continued the conservation of Spanish goats through the SVF Foundation and other breeders. The farm also left an endowment of stories about our experiences on the farm and on countless walks and horse rides through the surrounding forests.

These stories chronicle a moment in time on a small parcel of Appalachia. There have been, of course, many other moments on this land, containing many other stories. For thousands of years, our land was part of an old-growth forest. For many centuries, it witnessed native Americans hunting wild animals and harvesting seeds. The big trees fell to the first loggers about 150 years ago, about the time when the land was cleared of stumps to make its pastures. The surrounding trees grew again and have been cut again, several times over. Populations of deer were scarce, almost disappeared, and then surged. Fishers, once common, are now rare. Our community of birds is different from a decade ago. The forest, the animals, the mountain itself: they're all changing, and they all have stories to tell. We are current custodians, passing observers, and imperfect translators of their narratives. The stories in this book are not just ours but the land's as well. They provide a glimpse into the now and, in doing so, offer the future a bridge to the past.

Ken Liu, a novelist and futurist, observes that "We narrate our self into being, fill our soul with a breast-hoard of stories, and ultimately find solace in our spell-shroud as we decline into senescence." As we decline into our senescence, Priscilla and I hope to find solace in the shrouds of our storytelling. For now, the land itself still casts its spell on us through various incantations: bird songs, animal mutterings, wind whiffling through high grass. Its mysteries will continue for as long as the sun rises.

A Place to Love

Letter to Mom

Henry

Dear Mom,

Two decades ago—soon after your death—Priscilla and I went searching for something we couldn't define. Solace, isolation, adventure, a safe place? Maybe a place to love, like you loved your grandfather's farm near Cape Cod.

With restless souls, we headed for the mountains and, in our wayward searching, stumbled onto an old farm, a patch of Appalachia that seduced us into a relationship that's proved to be as unexpected, wonderful, and frustrating as any fierce passion.

We call our place Critton Creek Farm. It's halfway up a West Virginia hollow, along a meandering mountain road. Watch for our dirt driveway on the right, about a mile in, and head uphill. When the thick woods open to a house and barn, you'll be here.

I am writing this note to thank you for your spirit and your love. You worked many years, navigated hard losses and major disappointments, and made choices always with your children in mind. You approached life with skeptical pragmatism, a profound emotional reserve, and a deep pool of dignity and decency. You lived without artifice and remained honest to yourself all the way to the end.

Had you still been with us, you would've counseled against the farm's purchase: "Too impractical. It's in the middle of nowhere. The house is decrepit. The whole place needs work. Can't you find something more sensible?"

You'd have been right about the house, a quaint and dysfunctional seventy-year-old home resting unevenly on cedar stumps. Level lines and square corners went missing long before we stepped through its doors. The outhouse across the hill was larger than the indoor bathroom.

Insulation? Not so much. In winter, with its rusted gauges and an off-kilter fan, a kerosene-fueled heater wheezed like a twisted tin whistle blown by a weary old man. We eventually tore down the house and built a sensible one. Simple and solid.

Despite the problems, you would have understood the land's allure: forty acres of undulating pastures overgrown with briars and thorny roses; forty acres of untended forest, its high canopy covering scraggly rhododendrons, rocky run beds, and old logging paths cluttered with fallen trees; and a flat spot near the house, perfect for the garden.

From the start, the farm bred long to-do lists, like the ones you waved at me when I came home in the years before you died. Instead of lugging your old furniture from the attic or sweeping your backyard bricks, I have fences to install and fix, gardens to weed and harvest, wood to cut and haul, equipment to buy and repair, flowers to plant and pick, bird feeders to hang and fill, pastures to bush hog and drag, culverts to clean and re-set, barns to build and paint—and paint again.

"Face facts," you'd say. "These things have to be done. I told you it needed work."

Small failures sprout like transgressive briars, rooted in the necessities of making do when I don't have the right tools or enough experience. *The first woven-wire fence I installed wobbled like a sheet in the wind. I sharpen my saw's chain regularly, but it's always a bit dull.*

"Oh, stop complaining," you'd say, "You're privileged to have those problems." You were never tolerant of whining.

You might be surprised that Priscilla and I are still together. You knew from the start how different we were from each other. But we've managed to bridge those differences, sometimes with humor, sometimes with patience, sometimes with passion. You'd be happy that we're happy—at least most of the time. And I still love her as much as I did at our wedding.

"Well, don't ask for anything more than that," you'd say.

You would like our vegetable gardens. Of course, the weeds often win the battle; the back gives up and out; the heat wilts the spirit. (No complaints; just the facts.) At summer's end, we face the harvest's tyranny, those many hours needed to pick, sort, clean, freeze, and can. It's mostly Priscilla's burden, but I help when I can.

Like most gardeners, sometimes we have abundant tomatoes; sometimes, we lose them to early frost or late beetles. The raspberry harvest is almost always small. Those berries were your favorite, but then, you always felt a kinship with the rare and the lonely.

And the flower gardens, your special love, sprawl on the hill above the house and tuck in along the porches. They don their colors in the spring and change their dresses weekly. Early bluebells, crocus, and daffodils; mid-summer foxglove, cosmos, bee balm, and peonies; and a multitude of autumn sunchokes and marigolds. We designed the gardens with you in mind; each season, you come to us again in their glory.

A few years ago, you would have glimpsed our goat herd at its height. Fifty head of Spanish does, plus another eighty kids jumping around. *Critton Creek Farm became well known for breeding high-quality meat goats—thanks mostly to your daughter-in-law. All it took was a mix of work, love, and luck. No different from any small farm.*

Tragedy showed up, too. On a bitterly cold night, a barn burned, killing six of our best does and three kids. Ignited by a heat lamp too close to a hay pile, the fire left profound scars on the scorched earth as well as on our faith, hope, and memory.

Fifty does need their 200 hooves trimmed twice a year, and every kid needs shots during their first month. Managing the daily care, watching over the old and the pregnant, and witnessing the inevitable deaths of favored bucks became tougher as we grew older.

In the two decades since your death, I have walked many miles on this land, often with your spirit as companion and witness. Long before you passed away, you passed to me a resonant sense of place—a feeling that land matters, that it deserves respect, that we should improve it if we can. Your enduring love for your family's old farm and your ever-watchful spirit are resident here on this patch of land.

As I write this, the chill of midwinter is at hand. Last night, a bitter wind howled on the mountain's ridge. The trees' bare branches—their ruined choirs—are mostly songless; only the jays, perched in the nearby birch tree, squawk in the dimming afternoon sun.

I have a few years before I leave this place, before the sun disappears over the mountain for the last time. Whenever that departure comes, I hope to take my leave with as much grace as you lived your life. In the meantime, there's work to do: gardens to tend, pastures to mow, and fences to fix. This time, there'll be no wobble.

With much love,
Your restless son

Mud Between My Toes

Priscilla

The cool mud squeezed up between my toes. Looking at the sun on the water, I was mesmerized by the light glistening on the pond's smooth surface.

I was nine years old, exploring the farm that my dad had just bought, eleven miles outside Sylvania, Georgia. Buck Creek, to be exact, at the northern tip of the Okefenokee swamp. I was in the middle of farm country in the deep south.

It was a special day. Dad said he'd bought the farm yesterday and wanted to walk the fields. It was one of those rare occasions when he asked me to come along. Not my little brother. Just me. My whole body twitched with delight to be with him. When he was happy, spending time with him was the best. I thought he was magic.

Dad and I had one thing in common: a passion for horses. I'd loved horses since I was a baby. Dad loved them, too. And his dad, my grandfather, was also obsessed with them. Three generations crazy about horses. Walking the fields with my dad meant listening to his dreams about the farm and his horses.

As we crossed the field heading toward the pond, my thoughts kept going to the fact that our horses would live with us. On the farm! I'd go out the door and they'd be right there. I'd be able to ride them through the neighboring farms. I'd be with them all the time except when I was in school. That was my dream. And it had come true.

Dad and I pushed through the grass and finally reached the pond and stood on its bank, enjoying our first day on the farm. Dad sat down and began to take his shoes and socks off. I wore my flip-flops, my new flip-flops. Mom would have my hide if she knew I was walking through fields with just flip-flops, and it'd be worse if I brought them back muddy. I had to keep them nice for our visits to the beach. I reached down and slipped them off. Dad didn't notice anything about my flip-flops.

He'd left his shoes and socks in the grass and slid down the bank to the water's edge where he stood in the mud. I saw him smiling as he looked at his pond. His pond. He looked around at what he had bought and gave me a smile. "This is ours, Prissy," he announced. He wasn't talking just to me; he was talking to everyone in heaven.

The water was inviting—cool, slippery, soft. I gripped my flip-flops in a tight fist and held my arm way up. I couldn't risk letting them get dirty. I stepped into the water and felt the mud around my feet. The combination of soft sand and slick clay that oozed between my toes felt like jelly on creamy peanut butter.

Our aging Dachshund, Pepe, jumped through the tall grasses to follow us across the field. Now she was taking a swim, looking very happy and proud of herself.

Dad grinned as he enjoyed the cool, wet, squishy mud on his bare feet. We were having a good time. Dad, me, and Pepe. It was a rare occasion and one that I've treasured all my life. He was sharing this time with just me. Dad was alone with me.

As I gazed at the sun's light sparkling on the water's surface, I heard a noise from my father I'd never heard before. I turned toward him slowly because I didn't want to make any noise. His head was bowed, his hand cupped over his mouth, his shoulders hunched. Tears ran over his hand. He sounded as if he was choking. But he wasn't choking. He was crying. I had never seen him cry. I'd never seen any man cry. I'd only seen my little four-year-old brother cry—and other girls and a few women. What if he didn't want me to see him? What do I do?

I knew he was happy. He had bought the farm, which must have been his dream, so I was confused. My upbringing had taught me to just stand there and not ask questions. By the age of nine in my family, you learned to be quiet until you had a good idea what you should say and permission to say it.

He cried for a few minutes and then pulled a handkerchief from his pocket, blew his nose, and wiped his face. I tried not to look in his direction. I didn't know if he'd be mad if he noticed that I saw him. He put the handkerchief back, leapt up the bank, and retrieved his shoes and socks. He wiped his wet, muddy feet with his socks and put on the damp socks and his shoes. Partly from confusion and partly from fear, I stood still the whole time, still in the water, and said nothing. Had he forgotten I was even there?

Mustering up my courage to move, I climbed up the bank. I knelt in the grass, wiped my feet with my hands, and put on my flip-flops. I still wasn't sure if I should say anything, so I kept my mouth shut.

We finished walking around his new fields. He talked a little about where he wanted to put fencing. Being from the mountains of eastern Kentucky, he marveled at the wide flat fields, how easily a fence post went in. No shale! He smiled as his thoughts turned to things he wanted to accomplish on his farm. Dad was a wheeler-dealer. He also was a hard worker. Growing up in the mountains—and the army—drilled that into him. It also made him mean.

He never said anything about crying at the pond. I knew better than to bring it up, and I never did. Later in life, I understood those were his tears of happiness and sheer relief. He'd accomplished the most important thing to him at the time: buying his own farm.

Did he ever cry in front of my mother? If he did, were they tears of happiness? Did she hug him like she did her little boy, my brother? I didn't see much affection between my parents. There were happy pictures of them on trips with friends and sometimes they flirted with each other, but there wasn't a lot of happiness in our home life.

Dad kept the farm for years and he was content for a while. As a teenager, I realized that even though he was a gifted salesman and could turn on the charm when he needed to, he often felt alone and lonely. What I saw at the pond was a secret. As his daughter, I was raised to keep secrets and not ask questions. There were a lot of secrets in his house and a lot of questions that were never answered.

Flash forward forty years from the time with my dad at the pond to a different farm, a neglected but much-loved old farm, where Henry and I took a leap of faith. The house wasn't much to speak of, but the barn had potential. With some work, we could make it good for our horses.

Our first weekend on the farm as a family, the boys and dogs took off for the woods. Henry suggested we walk to the pond. I thought that was a grand idea and put on a new pair of boots I'd bought just for the farm.

We ambled down the road and cut into the big field, pushing through high grass, tangles of multiflora rose, and wild blackberry briers. Our pond sat in the middle of all that wild green growth. We found a grassy spot, took off our shoes and socks, rolled up our pants, and stepped into the water.

The cool mud squeezed up between my toes. Looking at the sun on the water, I was mesmerized by the light glistening on the pond's smooth surface. I was back at Buck Creek.

Henry rattled on about his dreams for our place. I pushed my feet into the mud a little more and watched the mud ooze up between my toes. Jelly on creamy peanut butter. The circle closed.

But not quite. This is my pond, our pond. Not my dad's. Tomorrow I can come and watch the sun's reflection scattered over the pond. And the day after, and the day after that. I don't have to wait for someone to ask me to come along. I go where I want to and when I want to. We'll keep no secrets on our farm. I don't have to keep secrets anymore.

I walked farther out into the water and stood perfectly still, the water cool and soft against my legs. My feet sank into the mud.

I looked toward the fields we had just walked through and back over my shoulder toward the mountain behind us. A giant weight lifted from my shoulders. Henry and I did this together. We'd come through good times and bad and sometimes I wasn't sure how we'd go on with each other. But this land, this keep of ours, promised a place of true solace. We'd fight, of course, but we'd be able to lick our wounds, and then love each other deeply and privately.

After that first day at the pond, I never saw my dad be emotional unless he was drunk. Maybe my parents were happy for a while, but they finally divorced. My dad wasn't very nice to my mother, who just lay down like a beaten dog.

Buying this farm took a leap of faith that Henry and I would be together for a long time. It was the right decision. We have stayed committed to each other, and to the marriage itself. We found our keep and it has always been a refuge. We trust it will always be so.

I stood in the water for some time, ignoring Henry, who still rattled on about what the place needed and how he was going to fix things. I'd been daydreaming, lost in my sense of being safe with my tribe and in our keep. I flashed back to my lime green flip-flops. I'd got them home and mom was never the wiser— they made it through two beach trips that summer. I chuckled, and then I felt the tears of happiness run down my cheeks as I watched my husband grin at me. With all the love I had ever felt for anyone or anything, I grinned back.

Decisions

Tiny Tim

Priscilla

Prelude

About three years after we bought the farm, I closed my clothing design and production business because rheumatoid arthritis left me exhausted and in constant pain. The loss of a vibrant professional career and the ravages of my condition brought my life to a standstill. I also started a complicated treatment that demanded a lot of rest for both my body and soul. Resting was hard. I walked in circles.

Two years after closing my business, I moved to the farm full-time, knowing it was a safe place where I could contemplate my future and examine my past. Long walks with the dogs, the solitude of the place, and a wonderful physician helped the healing process to begin. Like an animal licking its wounds in private, I cared for my body and tried to conjure up an understanding of the present without fearing the future.

This keep of ours gave me a place to heal and dream.

I read about traditional and heritage livestock of all kinds: Dexter cattle, Tamworth hogs, Spanish goats. One day, I found an old Hobby Farms magazine I'd brought home from a business trip years ago. On its cover was a Spanish goat. I read about small farms working to keep this very important heritage livestock in the breeding pool of today's meat goat industry.

I thought, "Well, I have a farm and I'm good with animals. I could raise Spanish goats. Maybe I could even do something important for our farming industry."

And so Henry and I went to goat school, a week-long workshop on how to breed and raise healthy goats. Later that year, we bought our first group of Spanish goats: five does and one buck. Eventually I had a core heard of fifty Spanish and Savanna does. At the farm's peak, more than eighty kids were born every spring. Tiny Tim was one of those kids.

Left for Dead

One of my pregnant does, Lu, was runty; she'd be a cull in November. She was a Savanna, a rare breed of goats that we had begun working with the year before. If I could get a good kid or two out of her, she wouldn't be a total loss. As a goat farmer trying to run a functional and profitable operation (well, if not profitable, at least not money guzzling), I had learned to be pragmatic. It was April, and kidding season was in full swing when I saw Lu giving birth in the front pasture; I decided to keep an eye on her.

I went about my work and periodically went back to check on her. By early afternoon, she'd had her kids. I walked down, hoping to see two strong doelings. As I got closer, the scene was sadder than I expected: two unimpressive bucklings. One was up, but the other was struggling. He couldn't stand. His legs wouldn't hold him.

I watched with pity as the little guy dragged himself through the grass with his front legs because his back legs weren't working. He was trying to get to his mom. She'd taken up with his brother, who had started nursing. It was hard to watch, but she was doing the right thing. She was tending to the stronger one—the one who had the best chance of survival.

"This is the way it's got to be," I said to myself. I believed—and still do—in good sound animal husbandry and a hands-off approach. I took care of my animals. But they had their job on the farm and I had mine.

I turned away. No one was there to get the gun, and I had feed to pick up in town. But I glanced back at the struggling little kid, still dragging himself along the ground toward a mother who didn't want him. I knew what chore would be waiting for me when I came home.

"He'll be dead by then," I said to myself. "Stick to your guns about what you will do and what you won't." I never bottle-fed kids. By this time, I was talking out loud. I walked out of the field, climbed in the truck, and left for town.

After I got back, I unloaded the feed and grabbed an empty sack for the body of the dead kid. Making my way through the field, I saw something in the middle, dragging itself in the direction of the barn. As I got closer, I was stunned. The little kid wasn't dead. Instead, he was using his front legs and all the strength he could muster to get to the barn.

I looked at him. "You shouldn't be alive," I said. "What do I do with you now?"

I stood watching him for another minute, then picked him up. He fell against my chest like a wet mophead. The afternoon wind was blowing and the shadows lengthening. It'd be much colder very soon. I stuck the kid inside my jacket, zipped it up for warmth, and walked through the pasture toward the barn, looking for his mom. Sure enough, the sorry doe was in there.

"At least she's taking care of the other kid," I said to myself. Although he wasn't the strongest buckling, he was decent size. I figured he'd be good for meat if he stayed healthy. As I watched the runty mother of these two sad kids, I made a mental note: this doe is definitely on the cull list.

The seemingly frail little kid I held had a strong heartbeat and was sucking on my sweater. Good sign. I wrapped him in a towel and laid him in the hay. Then I tied his mother to a stall wall and milked her.

A voice in my head spoke up. "You don't bottle-feed babies. Ever. If a doe can't feed her kid, the kid dies, and the mother is a cull. That's the rule."

But there I was—back in the house, making a bottle of his mother's colostrum. If he didn't drink it, he'd die. Despite my "hands-off" rule of farming, it just seemed wrong in this case, especially after all his work to get to this point.

When I returned to the barn, I held him in my lap and, after some struggle, he got the hang of the bottle. He downed it and finally perked up a bit. His head stopped bobbing, and he looked right at me, his eyes trying to find my face. He was tired and weak, but his belly was full.

I put him down, where he sank into the hay and made himself comfortable by poking the hay around with his nose. After working hard, he finally lay flat and fell asleep quickly. He was safe with his family, his runty family.

Then the dreaded question set in: Should I be doing this? I wouldn't have been standing there if someone had been around to use the gun. But it didn't play out that way that day. He was there, very much alive.

So, okay, I went against my own farming rules. "Where is the ice water in the veins, gal?" I asked myself. "Have I slipped up this time?"

But I had respect for this buckling, for his strong instinct to cling to life. He had dragged himself all afternoon across a cold and scrubby field trying to get to the barn and his mother. My instinct was to give him some help, not save him. He did the hard work: he stayed alive.

I looked down at Lu and her two kids. She raised her head, licked first one, then the other, and nosed them closer to keep them warm. "I can't believe it," I said to the three of them. "You're not a bad mother, Lu," I whispered to her.

More than Alive

The next morning, I thawed a bottle of goat milk that I'd saved from another doe and rushed to the barn, hoping the little fellow was still alive. Not only was he alive, but he was sort of sitting up in the hay. I picked him up. He felt like what he was: a long-legged, wobbly, newborn kid with a stronger neck than yesterday. He looked at me, and I blew gently into his mouth, imprinting my smell on him like his mother did when he was born. I wanted to be part of his family. I had a feeling he would need all the family he could get; life is not always easy in a large herd. "You'll remember me now, my little newborn buckling," I said and held him close.

Later, he learned to take a bottle in my lap, and I learned

to be patient feeding him. The question still nagged at me: What was I doing?

"I'm not breaking all my rules," I whispered to myself, "just some of them."

The bottle feeding went on for a couple of months and, even though it was more work, it was also enjoyable. After a while, he had started taking his bottle standing up, as if nursing from his mom. No more lap stuff. Then, after a few more days, I watched him gulp down the bottle of milk and then make his way to nurse a little more from his mother. One of his front legs was slightly shorter than the other, and he walked with an odd limp, but that boy had learned how to play the system.

One day, I arrived at the barn for his morning feeding, and he met me at the gate. He was having nothing to do with a bottle. He was a big buckling, and he was proving it.

After we knew he was going to live and be a small goat with a slight limp, a friend of ours named him Tiny Tim. Tiny for short. Perfect.

The guardian dogs soon came to understand he was a special case, and they grew protective of him. Once, as I came around the barn, I heard Tiny bellowing. Something was wrong. One of the guardian dogs, Bella, galloped past me. She knew something was wrong, too. We both turned the corner, and there was poor Tiny, up against the fence, getting walloped by four other bucklings who'd decided it was beat-up-on-Tiny day.

Bella ran right into the fray. By this time, Tiny was on the ground but still trying to fight back. Bella scattered the bucklings and started licking Tiny, who was trying to stand up, slinging his head around to fight his enemies and still bellowing in the loudest voice he could muster.

I scooped up Tiny and checked him over. No broken limbs—just some gashes on his side. I placed him in a stall with fresh hay, cleaned his gashes, and gave him some pain meds. Then, I found a small doeling he'd been friendly with and put her in the stall too. Tiny turned his head and bleated a defiant cry, as if to say to his tormentors, "I might be down, but I'm still here."

Tiny always needed extra help and different management than the other bucklings, and I had to accommodate his limitations. Tiny's legs didn't send him straight—he'd aim for a gate and miss on the left. I'd help him stay on track and, eventually, he'd get to his destination. He was never very strong, either. During the chaotic, jostling scrum at the trough at feeding time, Tiny couldn't stand up for himself. I taught him to walk around the corner of the barn so he could eat from his own bowl.

When the herd rotated pastures, Tiny couldn't keep up. I loaded him in the back of our farm truck along with the equipment needed for the next pasture. Tiny usually narrated his journey loudly, bellowing out comments to his friends running alongside the truck. They didn't notice, but it made him feel like an important part of the exciting pasture-rotation day.

Tiny grew to the size of a decent yearling buck, and then stopped growing. For the rest of his life, he looked like a nine-month-old kid. He remained a bit wobbly and was often a punching bag for the younger bucklings who'd just been weaned. Although the dogs tried to protect Tiny, the young bucks landed a few good licks each year. So, we castrated him, making him a wether in the goat world and allowing him to live with the does and doelings. His mother (who was lucky to avoid being culled for several years) seemed to enjoy his company in the pasture. At night, they bedded down together.

Uncle Tiny

The young doelings loved to play with Tiny. He was their Uncle Tiny and he had found an important role on the farm. Each year, when the doelings were weaned from their mothers, they'd weep and wail for days. But Tiny would step in and take care of them. He'd lead them to the pond, tell them when to get out of the sun, and make sure they napped in the pine thicket.

For four years, Tiny Tim was his odd little self. He enjoyed his job, giving many people a good belly-laugh when he looked at them with a mix of pride and befuddlement. Tiny sometimes wandered into the front yard with the dogs and barked along with them when a visitor arrived. He saw himself as part of the troop guarding the house. He was the hit attraction on the farm, and many folks have recounted their first meeting with this very unordinary goat.

A Tough Call

I had another firm rule on the farm: put animals down when they continue to carry a heavy load of intestinal parasites because they can pass those parasites to the whole herd. When Tiny was five, he developed a load of worms that he could never overcome, despite months of doctoring. He grew increasingly weak. I knew he had to be put down. It was one of those hard decisions the farming business requires. Sometimes you have to think with your head, not your heart. You can't go against the very reason you started farming.

Walking back to the house that evening, I said to myself that putting Tiny down was right, sound, and humane. It just hurt like hell. I wept all the way to the house and had no appetite for dinner. I just sat alone in the living room, watched the fire in the stove, and cried. If farming teaches you one thing, it's how to say goodbye. The next day Henry was home, and we put Tiny down. We buried Tiny next to some other special animals we'd lost over the years.

Tiny had a small body but a huge heart. He was a unique gift. He'd found a purpose in our world. And I learned something along the way: I learned to be careful about using the word "never."

All That the Land Contains

Henry

I claim no credit for the presence of most of the trees and creatures that now live on our farm. Nonetheless, I still feel responsible for what happens to them. But exactly how big is that responsibility?

In the fog of a day job and the constant demands of tending livestock, I had been oblivious to most of the trees and plants with which I share a place I deeply love. After finding other homes for our Spanish goats, Dexter cattle, and Tamworth hogs and leaving a demanding career, I began to pay attention to the creatures—animals, insects, bushes, and trees—that I had long ignored. How, I wondered, could I act responsibly to my place-mates if I didn't even know them?

I resolved to inventory the farm's inhabitants. Of course, identifying all of them on eighty acres would require several lifetimes. To make the task manageable, I decided to find a ten-foot by ten-foot plot within which I could list all the visible animals and plants that lived or walked there. If I noted what I found during the course of a year, I could track changes over time.

Such a strategy, I recognized, would provide an arbitrary and meager sample of my farm's residents, but it would start the larger task of understanding the possible effects of my future actions on the farm's landscape. In the preface to his 1986 book, *Arctic Dreams*, Barry Lopez, an American essayist and nature writer, makes this compelling observation: "In behaving respectfully toward all that the land contains, it is possible to imagine a stifling ignorance falling away." Close inspection of my chosen plot, I reasoned, would help define "all that the land contains" and take me one step closer to respectful meddling in its future. In the process, I might liberate myself from foggy obliviousness.

After considering several possible plots and selecting one that included bits of pasture and forest, as well as a young white oak, I got down to business one warm and misty April morning. Notebook in hand, I started with the grasses along the plot's southern edge. Thanks to Priscilla's careful tending over the past decade, our pastures contain a rich diversity of grasses including fescue, orchard, chicory, vetch, native long grasses, and brome, among others. I'm not a grass farmer, so I needed to enlist her to determine which kinds were actually in my ten-by-ten area. But she was busy that morning and couldn't help. Or wouldn't. She didn't think much of my inventory plan. In any event, I'd have to chart the grasses another day.

I moved farther in and knelt to scratch the loose soil. Several different types of worms wiggled in the morning light. I'm not a worm farmer either, so I took pictures, figuring I'd match them to pictures I'd find on the internet.

And then, stepping back, I looked at the oak: thirty feet tall with a broken limb, but otherwise healthy, and uncurling its first light-gold leaves. I'd include the tree itself in my inventory, but what about the birds that landed on its branches or the beetles burrowing into the hole left by the broken limb? My plot of flat ground suddenly became a vertical column. How high should it go? Eagle height?

I looked down. Brown needles from a nearby white pine lay scattered around the plot. Its roots almost certainly stretched underground, unseen and undisturbed. How far down should my now three-dimensional plot extend?

And then there was me. I had to include myself on the inventory because I was standing there, right in the middle of the plot. And, like almost every creature that walked or flew across that space, my presence made a difference to all the others. Learning about this small plot of land meant I was going to alter it in some way. Understanding and intervening were inextricably linked.

I slowly realized that listing all the species that intersected with even a tiny piece of my land amounted to an exercise in literalness. What would it accomplish, anyway? A list, however complete, is too concrete and static to provide answers to the mysteries that pester me. I want to know about inter-species interactions, intra-species relationships, seasonal contingencies, and energy exchanges—the way that life weaves itself together across one small spot. Nice idea, but beyond my reach. I'd need another long career, another lifetime to make true progress. As I grasped the futility of my plan, part of that stifling ignorance, that foggy obliviousness, fell away with a thud. It's complicated, really complicated, as I should have known. Believing in one grand tapestry of life may be inspirational, but pulling at its threads is inescapably humbling. On that misty morning, I tossed aside my notebook, stretched out under the oak tree, and gazed at the pale blue sky.

Through my back, I sensed the earth warming in the April sun. The ground hummed with a multitude of small movements. The earth was absorbing me—or, perhaps, I the earth. The longer I remained still, the blurrier the boundaries became.

Some crows cawed their gossip and circled a few times over an outstretched human. As they drifted away on warming updrafts, I imagined wandering the forest below them, strolling in silence and with sharpened senses. Damp earth, its scents promising springtime renewal; an almost invisible fawn, waiting for her mother's return; leaves uncurling, their first golds turning to shimmering greens; a shifty-eyed coyote padding with her kill over the murmuring run bed; warm air punctuated with cool draughts. I had seen and heard and felt these things before, but as a self distinctly apart from them. Now, the distinctions were vague, muddled, muted. I was in these things and they in me.

I'm not one to have out of body experiences or transcendent epiphanies, and my aging spine finally told me to stand up. Still, those moments left me with a different understanding of what my land contains and what it means to act responsibly.

It isn't so much what I do—catalogue the warblers or watch the jays, plow the meadow or mind the pond—as how I do it. I'm more deliberate now when I meddle with my land, aware that my understanding of the potential consequences always will be insufficient. The relationships among the trees, flowers, insects, and animals on my farm always will be more complicated than I can grasp.

Not knowing enough is no excuse to sit still. Reading and reasoning make my meddling deliberate—or so I hope. For example, if I'm going to plant native trees or flowers, I might as well start with the ones that host the greatest diversity of insects.

But science is an insufficient guide. Dreaming is needed too. When friends come and spend time on our land, they start to imagine things—and then to make suggestions. Some are economic: "You know, you could put in a Christmas tree farm." Some are purely aesthetic: "A batch of river birch would be beautiful in that low spot." And some are driven by a conservation impulse: "Plant more joe-pye weed all over; the caterpillars will devour it." Watching closely, I see them gazing over their imagined landscape, dreaming about what it will look like.

These well-meaning suggestions imply a substantial amount of actual work. Only a few friends come back to grab the shovel; even fewer are there to finish the job. Our farm is where dreams go to die of exhaustion. I know firsthand; the corpses of mine lie scattered round. But it doesn't matter. The land holds the memories of dreams that have passed on and the promise of ones yet to live as much as it contains its bees, ants, and oaks. Place-mates all.

Sid's Twins

Priscilla

What's my favorite thing about raising goats?

Kidding season in early spring! It was full of a little of everything on steroids, and I loved it.

Up with the rooster. Out the door as fast as possible to witness the night's events.

That year, the dogs in the back pasture barked almost every night. I figured the coyotes must be on the mountain behind the farm. Thinking about that made my heart beat like an electric drum, and I raced out the door. Are my dogs enough to stand up to the coyotes?

Don't forget the medical bag and extra ear tags.

Is it raining today? You forgot your rain gear yesterday!

Get the rain gear!

"Have a good day," I yelled to Henry as I headed for the pastures, barns, sheds, and pine thickets—anywhere I thought a momma goat would hide. Through the kitchen window, Henry gave me a good luck wave with a spatula while he made himself breakfast.

The weather that kidding season was nice—not too cold,

not too wet. I always worried about it being too hot, but we had mild weather and soft breezes that spring. Early in my life as a goat farmer, a first-time mom birthed a large kid in an extremely wet sinkhole in the middle of the hayfield. I just happened to be passing through the pasture on my rounds and walked over to see if she needed help. The kid couldn't get herself up, and the young doe wasn't sure what to do. The doe's status in the herd was low—she was a first-time mom, and her mom had been sold in the fall. Her first instinct was to keep the kid alive, and she licked the newborn to keep its circulation going but the kid needed to get out of the muddy rut she was in.

The problem was obvious. The kid needed to be picked up and put down somewhere dry and sheltered. I scooped up the kid as quickly as I could, carried her to the pine thicket, and laid her on soft, dry needles. The mother was worried—dancing beside me as I carried her baby—but she wasn't aggressive. I was happy about that.

As soon as I put the newborn down in the thicket, the young mother worked hard to lick and nudge the kid. Finally, it sat up, shaky at first, then strong. That inexperienced Spanish doe proved that the mothering instinct ran strong in her genes. Sound animal husbandry is careful watching, good judgment, and minimal intervention. That time it meant success for a new mom and a healthy kid. Boy, did I feel lucky!

As I walked around the pasture, I wondered if I'd need more luck this season. A pine thicket in one pasture provided good bedding and shelter through storms, and we had built small sheds around the other fields to give the goats protection and privacy. And there I was, roaming the pastures, looking for moms birthing babies, flashlight in one hand, tag gun ready to label the kids' ears in the other. But before you can tag, you must separate the newborn from its mother. Even though tagging is done as fast as humanly possible, once you poke something in a newborn's ear and it yells bloody murder, that mama goat will be on you!

Coming around the goat barn from the back pasture, I noticed a few pregnant does with an agitated gait and faraway looks on their faces. Birthing babies is long, hard, painful work. You could see the intensity in their eyes. These were seasoned moms who had their hiding places figured out and who understood what was happening. They knew my routine and weren't going to give me an inch. I'd be searching for them and their newborns later. I decided to continue my rounds.

A strong Spanish doe and second-time mom, Sid was obviously in labor, but things looked like they were proceeding naturally. She was hugely pregnant and walking around in circles about fifty feet from the barn. She probably would have twins. Many of the does having twins walk a lot. I saw no reason to worry. At least, not yet.

I walked on: out to the back fence, around the pond, back up the hill. Two does looked like they would probably give birth later in the day. No other action to take. The guardian dogs napped in the different pastures. I would hear them tonight, no doubt.

On my way back to the goat barn, I passed Sid again. She looked troubled: she walked as if she was drunk or in a horrific amount of pain, throwing her head back and forth and wailing in agony.

My body went cold and about a hundred things ran through my brain. I knew I needed to get her into a stall, and quickly. As my heart raced, I directed her into the barn using slow, exaggerated gestures so as not to scare her.

I drew closer and then spied the problem: two tiny noses were poking from her birth canal! One right on top of the other.

Twins were emerging at the same time, and they were stuck. My brain froze for what seemed like an hour, but I'm sure it couldn't have been more than a few seconds.

Sid struggled toward the open stall with the two tiny noses swaying. I remained quiet and deliberate as I guided her along, fighting my desire to run behind her and do something. But what? My heart was still racing. How could I move so slowly with my insides going a million miles a minute?

"I must get help," my mind kept repeating.

When Sid was safely in the stall, I put down clean hay and ran to the house.

I bolted upstairs to Henry's office, yelling out the problem in colorful language, I'm sure. I burst into his room. He was focused on his computer screen and talking on the phone— leading a conference call. He glanced at me and then back to the screen. Okay. My dilemma was not his to solve; this was my job. What I faced wasn't exactly in the job description of a goat farmer, but there I was.

I flew back down the stairs. I was going to have to do this alone and, for some crazy reason, I figured I could. I ran the whole way to the barn, which helped clear my brain.

As I entered the stall, Sid looked at me with complete trust. She didn't dodge or back away. The smell of clean hay surrounded us, and a gentle breeze flowed through the open barn doors. Although the light in the stall was dim, I could clearly see the tiny noses sticking out of Sid. A noiseless calm came over Sid and me. We understood the trust between us and the job that had to be done.

The kids' noses now stuck out farther—probably because of Sid's thrashing around. Maybe the kids were able to breathe. Maybe they were alive.

Maybe that was wishful thinking.

I stepped over to the water spigot and washed my hands as best as I could without soap. I couldn't remember where I put the plastic gloves, and there wasn't any time to look for them. Sid's breathing was heavy. She seemed ready. We got to work. I bent down behind her. She didn't move. Then, she gave a horrible bellow as the pain of another contraction came on. The pressure from the stuck twins was becoming unbearable.

"Well, this is it," I said to myself. And then to Sid, in a voice as calm and smooth as I could make it, "Hold on, girl. I've got to do this."

I pushed my hand and lower arm inside of Sid, sliding the kids back a bit into the birth canal. I worked my way down, trying to feel the babies' bodies and how they were positioned. Sid wobbled and screamed. The pressure of the two kids and my arm must have been intense. But she wasn't aggressive and didn't try to get away. Her back legs danced but never kicked out. With her front feet, she dug into the dirt, as if to brace herself for the ordeal. I kept probing with my hand and arm. I had to figure out how to reposition the kids because I was only going to have one chance.

Sid's head twisted with pain again as my arm moved inside her. I felt her contraction against my arm. I'd better decide because she wasn't going to last much longer. I slid my hand underneath the baby on the bottom, trying to make sure that both kids were in the proper position for birth: heads still forward, with front hoofs next to the head. I felt around the babies' necks for the cord but, as far as I could tell, it was out the way. The twins were in as natural a position for birth as I could manage to get them.

Finally, I pushed one kid back as gently and firmly as I could. It moved easily, and one nose disappeared. I brought my arm out slowly and held onto the head and shoulders of the other baby. Sid let out a huge yell as a contraction came. The baby moved, and suddenly the whole tiny head was outside. More yells and contractions, and the kid's shoulders were out. I moved back and let Sid finish the first birth.

She whirled around to check her newborn as I wiped its tiny nose with a soft cloth. The newborn shook and wiggled. Sid licked its face, checked the kid, and returned to giving birth to the second kid. With her head facing the ceiling, she gave out a yell and went into a contraction.

I became a sideline support for the second birth. "Just a couple more good contractions, girl. You can do this," I whispered.

Sid threw her head up and let out a deep yell. She bent down and pushed, her head twisting with pain. Finally, the second baby washed out in a rush and dropped onto the hay.

It hit the hay wiggling, glad to be free at last. I wiped mucus from its perfect face and gave it a fast once-over.

Sid turned feebly to nudge her twins and started licking them immediately, as all good mothers do with their kids. This birthing mother had been through an ordeal no mom wants to go through, but she never missed a beat taking care of her newborns.

Time passed quickly. I shuddered at the memory of moving the kids inside of Sid. I looked down at my arm, covered with dried blood. It was shaking.

Both kids were alive and acting just like two newborn kids should: trying to stand, shake, and move around, falling, getting back up, and trying to find the needed equipment on mom. Sid had birthed two big, healthy doelings. I couldn't have been happier.

Sid worked the twins to her teats, licking as she herded them to their first meal. She moved gingerly but with great confidence around the doelings. Finally, they settled against her full bags and began nursing. Now and then her head would twist and she would grimace in pain, but she never slowed the mothering of her twins. With the kids having nursed and ready for a nap, Sid felt comfortable enough to lie down and curl around her newborns. Stiff from all her labor and sighing from exhaustion, she closed her eyes.

Looking at the three goats resting in the hay, I was overcome with awe, relief, and wonder for these magnificent animals that had the ability to go through this birth. And it turned out okay! They were all okay! Even I was okay!

Now I understood what my farming friends meant when I asked them about a situation like this one. What should I do? They would smile at me and say, "You'll figure it out—or you won't."

After I saw that Sid and her kids were fine, I made my way to the feed room to make some special food—a mixture I developed for does after birthing, especially for gals who had hard times. Grains, beet pulp, minerals, electrolytes, molasses, warm water, and baby aspirin for pain and inflammation. Sid gulped it down, then went on to some fresh water. After a long drink, she dropped her head into a bunch of alfalfa. She still moved slowly and with some effort, so I decided she would stay in the barn for the night and snuggle in the hay with her twins, who were already balled up in the hay together. I'd see how they were in the morning.

My legs felt limp and my arms were numb. I had been running on instinct, adrenaline, and hopefully the right judgment. Now I was just plain wrung out. I sat on a bale of hay to watch the new family.

"Hello?" Henry's voice came from the front of the barn.

"Come on back here," I called, standing up.

"I'm sorry I couldn't get here sooner—"

"Sid was such a trouper! You should've seen her!"

As usual, when we were excited, we talked at the same time.

"We are all fine," I said with a relieved smile.

His face mirrored mine.

We stood for a while in the dimming light of early evening and took in the amazing sight. The barn was warm and quiet. The sweet smell of hay and molasses from her mash helped me relax as I leaned against Henry. I still couldn't believe they were alright, but they were.

The next morning, the twins were fine, but Sid still moved stiffly and slowly in the stall. She could use more time with just her twins rather than being out with the herd. I fashioned a small corral for them near the barn. They would still be separate from the herd, but next to their pasture and able to rest in the stall when they wanted.

The herd gathered along the fence to see the newborns. Sid visited with the goats in her family group, most likely telling of her ordeal and the way she pulled it off with a little help from the farmer woman. She was the star on the farm for a while.

She stayed a star in my book for quite a while.

Mindful Meddling

Henry

I looked casually through my gunscope, sighting along the tree line in the unlikely case that a deer would appear before the sun set. It was late in the season and many hunters had walked the mountain above me; in response, the deer had become mostly nocturnal. My spot—a stool in an old shed with a missing board on its backside—provided a narrow view of a field where deer bedded down at night.

I brought the crosshairs to rest momentarily on a low branch of a Virginia pine. I was restless, thinking of projects clamoring for my attention: fence boards to fix, tractor oil to change, bills to pay. "Sitting here is a waste of time," I grumbled as I admired the flowing curve of the pine's long, swooping limb.

And then, as if I were in a silent dream, a doe walked into the scope's circular eye.

I didn't see her coming. She was right there, still walking. I rotated the rifle slightly, keeping her in my sights. She moved quickly, then slowed. A big one. Graceful. Elegant. She stopped and raised her head, her attention captured by two horses running in a nearby pasture.

I exhaled and pulled the trigger. Flash, knock-back, piercing boom, soft echo, silence. The doe jerked around and sprinted to the tree line. As she headed down a steep slope toward a run bed, her white tail flicked once, twice, and then she was gone from sight. I waited a bit before I walked out to find her.

I don't consider myself a hunter. I hunt, yes, but I'm ill-suited to the task. I prefer not to wait hours in the cold for wild animals to walk by. The scenery is always beautiful, but eventually my mind wanders to other projects. My aim is accurate most of the time, but I find little pleasure in the physical act of shooting. I don't like loud noises. The taking of a life doesn't come naturally. Or so I like to think.

Our West Virginia farm faces a problem familiar to rural and suburban families: deer are too numerous because they have few predators. Many communities argue about the deer. Allow them to become wild pets? Bring in hunters to reduce the herd? Inject birth control hormones? These questions won't be settled soon or easily.

For me, deer were once emissaries from the wild, representatives of a nearby world that included bobcats, coyotes, fishers, and bears. Recognizing that we share this land, I paid them respect. But my feelings have changed. By themselves, deer now are ever-present threats to gardens, fruit trees, and the forest itself. The wild orchids, lilies, and young oaks that once dotted the forest floor have disappeared. In the understory on the mountainside, only the wild blackberries still thrive. All around us, deer mangle gardens, eat the fruits of carefully tended orchards, make forest floors barren, introduce invasive parasites and disease, and contribute to road accidents. They have become representatives of nature's distress, their population growth signaling a world tilting on the edge of a troubled future.

Last fall, I seeded an acre of meadow with wildflowers to support and diversify the farm's pollinator population. It sits far back from the house at the boundary between cultivated land and untended forest. I was worried that the deer herd would eat every stem that sprouted in the spring, long before the flowers bloomed.

What to do? I could fence the acre of meadow, but that's not financially feasible. I could use various chemical products, but they can have unexpected side effects on other animals and insects. I could string wires festooned with shimmering tape to scare them away, but they would learn quickly to ignore the flashing. I could reduce the herd through out-of-season harvesting, which would require crop-damage permits from the county. But killing even a half dozen deer would hardly make a difference to the herds around us.

Reducing the deer population or preventing them from eating in their own front yard invites a battle with nature—a battle on land whose ownership is in dispute. Is the meadow mine because I've met all the requirements that human society has established to give me rights to the claim of ownership? Or is this land actually still one of nature's parcels?

Some would argue that I should be content simply to watch nature take its course. Make no effort to support pollinators, who themselves face serious threats to their survival. Plant no new hickories to increase the diversity of hardwoods. Remove none of the hundreds of Virginia pines to give the white pine saplings a better chance to grow. Cut no dead trees to burn as firewood in the living room stove, offsetting the propane that would otherwise heat the house.

But to choose inaction ignores the reality of my presence. For better or worse, I'm part of the action. I walk the land often and I work the land—or, more accurately, collaborate with the land—by keeping the pastures healthy for growing hay. It's my nature to intervene, sometimes in the service of my own esthetic needs, sometimes in the service of my family's needs, and sometimes with the hope that my interventions will be more beneficial than destructive.

I spotted the doe within a few minutes. She'd collapsed under a large white oak, halfway up the far side of the run bed. I needed my tractor to pull her to a level spot along the tree line. Field dressing a deer is always messy; no exception this time. Stomach, intestines, liver, heart, and spleen spilled out in a rush, flopping to the ground with a sloppy squish. Her blood's dank, musty scent gathered close, drifting downhill as the evening deepened. The coyotes would enjoy a generous dinner, and the crows would clean up before tomorrow's sunset.

I've gotten to where I can be dispassionate about the butchering business. Neither attractive nor odious, neither celebratory nor grim, quartering a deer is just rough justice on a farm that borders the wild. I'll not ask someone else to do it, and we'll eat the meat as the gift it is.

Still, it's difficult to reconcile life's horror with my desire to live a compassionate life. The deep paradox of this farm is visible in every undulation of its hills, in every turn of its forested paths. No spot has not witnessed the brutality and randomness of death in the natural world; no spot is distant from breathtaking beauty. No spot has not been meddled with by human hands; no spot remains beyond nature's grasp. And every choice I make to meddle with this land, however well reasoned, has consequences I can't foresee, no matter how thoughtful I might be.

What to do with this dilemma? Why hunt when I'm not a hunter? For me, the answer is to tread lightly and remain aware of the contradictions and tensions of consciousness itself. I'm a product of the randomness of evolution as much as any creature, and yet, only as a human can I write this essay. I'm both apart from and a part of the natural world. Okay. I'll straddle that border and lean, as much as I can, into the light on both sides.

I'll stay close to the land, watching its natural habitats: the forest's understory, hoping to catch a glimpse of the few orchids the deer overlook; the pond, which serves as the local pub for a retinue of regulars; the abandoned road with its recent settlers, including a large family of sassafras. Each place has its own sounds, smells, and visitors—its own life. And yet, the more familiar I become with those places, the more I feel the tension between watching and doing. I'll want to intervene. Why not enlarge the pond a bit to accommodate more migratory ducks and geese? They certainly need help these days. Why not plant some native oaks on the old road to prevent more erosion and support the caterpillars? Of course, I'll have to protect the saplings from the deer.

I'm trying to make this farm an honorable place by way of mindful meddling. I'm tempted to ignore the complexities of my possible choices and just do what "feels right." But I won't abandon the effort to anticipate the consequences of my actions. I'd rather struggle with powerful contradictions than live with the pretense of simplicity.

The Dilemma of Loving Hogs

Priscilla

His eyes were not evil, but they spoke of anger, boredom, and despair. The pupils were dark, but not black; the bright reddish brown around them accentuated their rage. He snarled at me with a growl from deep within his throat.

If he could have escaped from his pen, he would have killed me. He would have killed anyone. He knew what was happening to him. And he was pissed.

The only other hog in the pen was resigned to his fate. He paced slightly but mostly just waited in the corner, watching. He could become aggressive too, like his companion. Some days they beat up on each other. Overcrowding does that to creatures who need some dignity.

Because he was somewhat calmer, the quieter hog would yield better meat. How many generations would it take to breed for a calm personality? Two? Three? Breeders like me think like that. The angry hog will have tough roasts because he's stressed. The rest of the cuts probably would be tough, too.

Is this standing back and being a businessperson my coping mechanism? Part of who I am? Where was the balance I wanted to keep?

The hog pen was at my neighbor's farm. About six months ago, we'd bought four feeder hogs—two for him and two for me. We agreed that he'd keep them at his place because he'd raised many hogs before, and I hadn't raised any. He'd do most of the caregiving and I'd do my part: paying a monthly amount for his work, covering my share of the vet bills, and bringing over food scraps like chicken carcasses and old veggies.

Right from the start, I was uncomfortable with the size of the hog pen, but at least it was in the shade. I was the greenhorn, so I kept my mouth shut.

But the piglets had grown up and space was tight. They weighed about 350 pounds now and were ready for the butcher, which is why I was there. Lured by some apples and chicken bones with meat still on them, the hogs walked into a cage on my pickup truck. I shut the cage door and closed the tailgate.

They finished off the bones and turned around and around in the cage's thick layer of soft hay. I climbed into the truck and headed out. After a few miles, I saw in my rearview mirror they were both lying down.

The butcher was an hour away and I spent much of the time thinking about how my neighbor and his wife had dealt with this experience. She and I had worked on a project for the community and had enjoyed our time together. She knew I wanted to learn about farming and so her husband offered to help me with my first batch of feeder hogs. Both were generous with their time, and I appreciated their friendship. I'd learned a lot.

But I was an outsider. Some of their family and friends probably questioned why they had helped me out—especially because I was known for spouting off about humane treatment of animals and more progressive ways of managing pastures and livestock.

Our part of rural West Virginia is not known for its broad-minded approach to farming. I try to stand up for my principles without a heavy hand or a judgmental heart, but sometimes it's hard. One day I discovered that a nearby farmer was treating his animals in a way that was against everything I believed. *He's not evil,* I thought to myself. *He's just part of this culture. The local folk think highly of his family. They must be decent.*

But when do you take a stand that's different from the customary way of doing things and the beliefs of decent people?

Even though I'd never raised feeder hogs, this wasn't the first time I'd been involved in butchering them. When I was a little girl, my parents often took me to my Mamaw's farm in the mountains of eastern Kentucky for Thanksgiving. We timed the trip so dad could go hog hunting with his relatives.

All the farmers in the hollow let their hogs roam through everyone's woods and clean up cornfields after the harvest. Each farmer notched their hogs' ears, so they all knew whose hog was whose. Depending on the age of the litters and the weather, the hogs wandered the woods from summer to winter. The farmers would tell each other when they caught sight of the roving porcine band.

In late fall, around the holidays, Dad's family came home to visit. Everyone participated in hunting the hogs, butchering them, drinking homemade moonshine, and telling stories. And everyone made sure to mark their hams and bacon before hanging them in the smokehouse. I loved the fresh ham sandwiches we ate as we drove home to Georgia.

I smiled an old, wrinkled smile and pictured myself as a little girl, following Mamaw to the barn through the new snow to feed the piglets the kitchen scraps. The piglets would go out to the fields and woods when the weather was warmer. I always thought their life was the best, roaming the mountains.

I checked the rearview mirror. The hogs were still asleep. I was glad they were having a good nap, but I was bothered that I hadn't raised them on my own farm, according to my own standards. I wasn't the little girl in the back seat of the car coming back from Mamaw's. I was running the farm, and it was my job to do what was best for the farm and its livestock.

As I drove, I resolved again to match my farm management with my beliefs. I'll do things differently from the folks around me. I won't say I want dignity for my animals and not give it to them. No hypocrisy on my farm. I decided that we'd house all our livestock on our farm, from then on. We'd only work with the animals we raised. If we didn't have enough pasture, then we wouldn't add animals.

I finally arrived at the butcher, where I'd been taking my goats for several years. I always enjoyed seeing the group of kind professionals who worked there.

"Well, hi, Ms. Critton Creek Farm. I see you've brought us a couple hogs," Sara said with her usual lovely smile.

"Yes, Sara, I've got a couple of hogs," I smiled back at her. "Did I fill out my cut sheet right this time? You know, next year, I think I'm going to try pasturing a few hogs."

I pulled the truck around to the chute that led to the kill room. With help from one of the butcher's assistants, I positioned the door of the cage in front of the chute, turned off the truck, and hopped out. As I walked along the side of the cage, I saw trouble brewing. I opened the cage door, and the calmer hog came right out—he had been waiting to escape from his mean traveling companion. The other hog stared at me like a gladiator staring down a lion. This guy was not going to budge. He knew he had everything to lose.

I started to climb into the cage. The fellow helping me grabbed my arm and said in a thick Spanish accent, "No! Lady! No!"

I looked again at the hog's eyes. They had gone cold. Totally cold. He was not just pissed; he would have hurt me if I had gotten in there with him. He was smart enough to know his fate and not powerful enough to stop it, not powerful enough to craft a way to help humans understand that his time on earth could have been better, kinder. He could have been kinder if his humans had been kinder to him.

The fellow helping me brought out a cattle prod and offered it to me. He couldn't do anything to get the hog out of my truck—the butcher's insurance requires farmers to remove animals from the truck and into the chute before the butcher can take over.

I banged again on the side of the cage. The hog glared at me. I banged once more. This time he charged the side of the cage. The thump was frightening.

The fellow tried to hand me the cattle prod again as he glanced behind us. Other farmers had arrived to drop off livestock and their trucks were waiting in line behind me. I had to get the hog out of the cage.

I held the prod, turned on the electricity, and struck the hog. He squealed in pain and surprise. I did it again and he moved closer to the open door. And then, he whirled around and glared at me with eyes that were now full of hatred.

I struck him two more times in the same place. He let out a terrible yell and ran at the cage one more time, denting the bars. He finally charged through the open cage door, roaring like a lion, and flinging slobber as he waved his huge head in anger. He ran down the track and into the kill room, finally giving in to his fate.

I grabbed the truck, shaking, still clutching the cattle prod. I was dazed but eventually managed to close the cage door and the tailgate. I thanked the man for helping me, gave him back the cattle prod, and apologized to the farmer behind me for making him wait.

"No problem," he replied, his eyes sympathetic. Maybe I was still shaking. Maybe he had been through the same problem.

I drove home, but my mind was mush. I listened to the news that I didn't care about and tried to remember if I had taken anything out of the freezer for dinner. I didn't want to think about the hogs, or the mistakes I'd made, but anger and frustration welled up. I pulled to the side of the road and started beating on my steering wheel. I yelled at the top of my voice until my throat was sore. I was angry at the whole mess: at how those hogs had been treated, at the horrible system of food production, at myself. I remembered what Temple Grandin, an American academic and farm animal behaviorist, once wrote: "I think using animals for food is an ethical thing to do, but we've got to do it right. We've got to give those animals a decent life and we've got to give them a painless death. We owe the animal respect. If we lose respect for animals, we also lose respect for ourselves. It's as simple as that."

So what if I sold some pork and maybe broke even? So what if my family ate good chops? Was this whole journey worth it?

At home I took refuge in my new *Stockman Grass Farmer* magazine, which included an article on pastured hogs by Joel Salatin, a knowledgeable farmer and the owner of Polyface Farms. Experience had taught him that putting hogs in a pasture would yield excellent meat and happier animals. I could do that!

I discussed my ideas with Henry and the next spring we fenced in a half acre of land with electric rope. It was a nice spot for hogs because it was overgrown with bushes but still had quite a few oaks, maples, and pines.

After a bit of searching, I found a farmer selling Tamworth piglets—a heritage breed. The sow had been pasture raised and was keeping her litter in a big barn with a small paddock. The farmer believed the sow and her litter were healthier and happier together. I agreed as I watched them move around the barn in a natural, contented way. The sow was calm, the piglets playful—hilarious in their antics. I bought my four first piglets and brought them back to their own half acre. Their lives might not be long, but they would be good lives. I'd make sure of that.

The piglets learned to respect the electric fence and enjoyed their home. They had a shed they loved. At various times, they slept on the shed, in the shed, and under the shed. There were autumn olive bushes to lie under and munch on, and they could roam among the trees, rooting for grubs.

When the pigs were at the right weight, Henry brought them to the butcher because he had to use the truck for some other job. He said they did fine. He let them take their time and they played follow the leader down the chute.

The next spring it was time for a new batch of piglets. This time we bought Tamworth and Berkshire crosses: three girls (whom we named Iris, Rosie, and White Foot) and one boy (Indie, short for Independent; he had to be independent because the girls were always bossing him around).

The piglets were shy when they first arrived, retreating to one corner of the pasture whenever I showed up. Eventually they became used to me, and I enjoyed their attention. I don't know if they enjoyed mine.

Near to the hogs' half acre was a small overgrown field with a wet area where water puddled up from an underground spring. If the hogs could root around there, maybe I could turn the wet spot into a seasonal stream, which then would make the field a good place for the goats. We moved the pigs into this field and, sure enough, they began rooting all around the stream. After a few days of demolition work, they had themselves a small wallowing pond and had widened the stream bed. The water from the spring started to flow more consistently and eventually, after the pigs left, it provided a great water source for the goats.

A day with hogs is never boring. The pigs followed me around the pasture and loved it when I played lifeguard and watched them wallow in their new little pond. It was the farm equivalent of watching your kids at the pool. I was fond of each of them, and they weren't afraid of me. The job of pasturing hogs was a joy and I never thought of it as work. Their curiosity and boundless energy for everyday adventures always put a smile on my face.

One time a goat got her head caught in the fence and it was one of the hogs, Indie, that ran to Henry to come and save their friend. The other hogs stayed with the goat to make sure she was safe. After Henry freed the goat's head, the four hogs danced their way to their dinner.

When it was time to take them to the butcher, I told Henry I'd do it. I took the goats and steers so why not the hogs?

During the week before they were scheduled for the butcher, I fed the hogs in the horse trailer, which I'd parked in their pasture with the back door open. After every meal, they loved sleeping in the deep piles of hay I had spread on the floor. The morning they were going to the butcher, I fed them as usual and then closed the door.

I was working with a new butcher and his policy meant that I'd pull the truck around back and off-load the hogs before stopping by the office.

A worker met me at the gate. The stall for the hogs was down a center hallway. I opened the back door of the horse trailer. The four hogs peered out and took a few steps down the trailer's ramp. "Come on," I said in a quiet voice, and I patted one on the back. They followed me to their designated stall, and we all walked in. It was clean and there were some other hogs in the next stall. The three girls ran over, hoping to meet some new friends. But Indie stood beside me. He had always been the shyest of the four. He looked up at me and didn't move.

"Are these pets?" asked the worker.

I wasn't sure what to say. I stepped out of the stall and stood for a minute. The worker closed the gate.

Indie stood near the gate, still looking at me, puzzled. "Why was I leaving them here?" was written all over his face.

I turned and walked out. I pulled my truck and trailer around to the front, went inside, filled out my paperwork, tried to be pleasant, and left.

I cried all the way home. This experiment wasn't working. The hogs were too smart, too personable. I loved their intelligence and gentleness. I treasured their sense of humor and playfulness. I couldn't keep a safe distance from my feelings.

In the eyes of that first hog, I saw anger and desperation. I never wanted to see those eyes in any of my animals. He had been stripped of his humanity. Humans had let him down.

In Indie's eyes I saw trust and confusion. That was worse.

Surprises

Surprises

Of These Mountains

Henry

Stopping By

On the day after we bought the farm, Priscilla watched from the porch as a faded red truck motored slowly up the driveway. Two men sat in the front seat, looking fierce with their camo hats, unkempt hair, unshaven cheeks, and unsmiling faces.

"Oh, Lord," she whispered as she remembered the men on her father's side who lived in Bear Hollow, Kentucky. "This could be trouble."

The truck stopped in front of the porch and the driver rolled down his window. He turned to look at her.

"You Miz Ireys?"

"Yes," Priscilla answered. "Who are you?"

"I'm Eddy. Eddy Shirley." He turned his head forward, his wrist resting over the steering wheel.

Priscilla paused, waiting for him to say something else, or at least introduce the other fellow. Eddy just kept looking straight ahead.

"Hello, Eddy," Priscilla said. "Would you like some coffee?"

"Yeah."

Eddy slipped out of the truck and walked up the porch's few steps.

"Your buddy can have some, too," Priscilla said.

"He's my nephew." Eddy turned around. "Come on, Chris," he hollered at the truck.

Sitting at the kitchen table with the two men, Priscilla chatted about why we bought the place, where our children were in school, and how the summer seemed unusually dry. Eddy didn't say much at first, but then he warmed up. Turns out he was full of news and gossip from the hollow. He lived up the road about a mile, close to where he was born. He talked about how little it had rained, how our neighbor made snowshoes, and how his aunt had died in a fire at the homeplace. He'd worked off and on for the fellow who'd sold us the farm: mowing the lawn, helping to build the woodshed, and doing odd jobs here and there. Chris added extra sugar to his coffee, drank it quickly, and never spoke a word.

"Well," Priscilla said after an hour or so, "I've got to get going."

"I'm out of here," Eddy announced. He stood up and headed out the door. His nephew followed him, silent as ever.

It was the start of a long relationship. Eddy became part of the farm quickly and, over the years, helped us dig trenches, build sheds, and care for our livestock. He was at home in the forest, too, because he'd lived on the mountain all his life. He knew its contours as intimately as the deer, coyotes, and bobcats who roamed its run beds.

Eddy built his cabin by himself decades ago on a few acres of his parents' land. Brothers and nephews often stop by his house, especially in hunting season. They gather there to fix trucks, shoot pool, sight in their guns and bows, or have a few beers. Mostly he lives alone, but not entirely so. He throws corn out for deer, turkeys, and squirrels. He's watched them for so long, he knows their habits and moods. If he walks to his garage, they look up but don't pay him much mind. During hunting season, he may shoot them in the forest but never in his yard. They're safe there. In many ways, they are his companions. To them, he's just part of their mountain.

Building Fence

My first major project after we bought the farm involved building a post-and-board fence to enclose a horse pasture. I had never built a fence—I'd grown up in New York City and spent a lot of my life wandering libraries, staring at computer screens, and listening to people lecture—but it seemed straightforward: put in the five-inch-round, eight-foot-tall posts eight feet apart and three feet down; nail three rows of five-quarter boards to the posts; and hang a couple of gates. A local sawmill delivered the boards, and I bought the posts from a nearby hardware store.

To dig the post holes, I rented a Bobcat with an auger, which is like a large corkscrew. I'd never used a Bobcat before, but it seemed straightforward: one lever to raise and lower the auger; one lever to adjust its speed.

I hired Eddy and Glendon, another one of his nephews, to help. With everything at hand, we started early one morning. The seemingly straightforward immediately became the truly difficult.

I'd assumed that placing the posts in a straight line would

be easy: we'd just have to eyeball it. Not so. After drilling nine or ten holes, we checked the placement. Far from an orderly line, the holes swerved this way and that.

"We should have used a guide string," I admitted, as we started to re-drill some of the post holes.

"I thought we might," Eddy said.

Thanks, Eddy, I thought. *How come you didn't say something earlier?* Over the years—at least when it came to farm work—Eddy rarely failed to let me make my own mistakes.

Digging the post holes at the bottom of the pasture was easy: set the auger point on the ground, start slow, keep the pressure even, increase the speed, reverse to pull the drill out, repeat a couple of times to clean the hole. As we moved up the hill, the topsoil thinned, and the auger struck shale a foot or two down.

"Bring it down hard. It ought to break the shale," Eddy suggested. I raised the auger above the hole and drove it down. When it hit, the Bobcat leapt off the ground, nearly throwing me out of the cab.

Eddy poked at the rock with the digging bar. "It hain't broke," he said.

I tried again, this time with my seatbelt attached. The Bobcat bounced again. No change.

"Try here," Eddy said, moving to a spot two feet up the hill. I was sure we'd hit the shale shelf again and we did, but this time the shale cracked after a few blows from the auger, and we dug down almost three feet.

As we moved along, the hill became steeper. At one point, I raised the augur high while driving parallel to the hill's slope. The Bobcat tilted precariously. A slight panic spread from my stomach to my face. Eddy must have finally understood that I didn't know what I was doing. He never laughed, never made fun of me (at least not where I could see), and never suggested that I stop for my own sake. Instead, he hollered from where he was standing. "Don't lift it so high. Turn straight up or down if you feel it going." Simple, obvious suggestions. They might have saved my life.

It took us two days to finish digging the holes and setting the posts. Because we had to dodge the rock shelves, some posts were ten feet apart rather than the standard eight feet. The boards were all about eight feet long. I'd need to go back to the sawmill for some ten-footers.

But Eddy, who'd built many fences in his life, didn't see the situation as a problem; it was just part of fence building. He cut a three-foot-long section from some discarded boards, nailed it to an eight-foot board with a one-foot overlap, and there it was—a makeshift ten-footer. After we'd nailed the ten-footers to the posts, he set a perpendicular two-by-four post on the ground and nailed it across the fence's three rows of overlapping boards.

"That's stout," he said, mostly to himself.

I agreed it was strong enough for a horse pasture, but the aesthetics were slightly off. It looked makeshift because it was makeshift. Priscilla didn't think much of it either. But, as the years passed, no animal ever challenged that part of the fence. It not only endured, it prevailed.

Finding a Tractor

One day, less than a year after we bought the farm, Eddy drove into the yard, stopped near me, and rolled down his window. "You might want to go see this tractor."

Priscilla had spoken to Eddy about needing a tractor for

the farm. They'd agreed we wouldn't be able to manage the work without one.

"What tractor?" I asked.

"Over there in High View."

"Where's that?"

"Off 50."

"Route 50?" I replied. "Where off 50?"

"Oh, you know," he said, nodding in some vague direction, "where that store was they used to sell pretty much everything. Right there near the line 'tween Virginia."

"Okay, Eddy," I said, giving up. "Get in my truck. Let's head that way and you can tell me where to turn."

We rode for about forty-five minutes to a small town near the Virginia border where a farm dealership was selling a used thirty-seven-horsepower Kubota tractor with a backhoe. Eddy had driven past it earlier in the day and thought it was a good deal.

"Won't last long at that price," he said after we'd started it up. "Sounds good. It's not got the hours on it."

The problem was, at that point, I had no experience with tractors. None. We had bought the place just a few months previously. I was in unfamiliar territory. I'd never driven a tractor like the one I was looking at. I had some idea about how we'd use it on the farm—digging holes, moving dirt, pulling things—but couldn't envision our specific needs.

I hesitated. This tractor would be a big investment. What was I doing there at all? A city kid standing next to a loudly rumbling tractor whose rear wheel was almost as tall as I was. In the cartoon version of this story, the next frame would show a tractor chasing a man downhill.

Eddy stood nearby, expressionless as usual. He'd said what he thought. No need to say more.

"You think I can manage this?" I asked him.

He paused and looked away. "Yeah, I'd say you'll probably get used to it." From Eddy, that was a strong vote of confidence. So Eddy found us a tractor, and we've been grateful ever since. He was right about us getting used to it. Priscilla and I have spent many hours on that tractor: digging holes, spreading manure, mowing fields, dragging logs, pushing stumps, hauling wagons, scraping roads, and so on and on. It's become part of the farm, too, both substance and symbol.

Bringing Down the House

Five years after we bought the farm, we decided to live there full-time. That meant we had to replace the seventy-year-old, two-story farmhouse because it wasn't worth renovating. To tear it down, I rented a fifty-horsepower tractor with a large backhoe—a much larger machine than the tractor we'd been using since our first year on the farm. To begin the demolition, I pulled the porch off the house and then systematically clawed away the outside walls of the first floor. With Eddy, Nick (my elder son), and Joe (our neighbor from up the hollow), I worked carefully to save as much of the old rough-cut lumber and tongue-and-groove floorboards as possible. Eventually, we opened each side of the house until we could see straight through from all sides.

Despite the absence of the first floor's outside walls, the second floor remained intact, held aloft only by a large center post, as if a waiter were balancing a tray with one finger. The obvious strategy was to place a chain around that post and pull it down with the tractor. But none of us wanted to walk into a house with a teetering second floor. It could pancake without warning.

It just so happened that a fifty-foot Virginia pine was leaning over the house. The tree, with a trunk twisted like a taut muscle, was well past its prime and beginning to lose its top branches. Why not cut the tree so it would fall on the roof and knock the second story flat? It was the opportunity of a lifetime; none of us would ever again force a tree to smash a house.

I glanced at Eddy. He'd been logging since he was a teenager working with his dad at their sawmill and knew the way that trees with twisted trunks can spin and snap in odd directions. He was experienced enough to respect the power of heartwood.

"How's it look to you, Eddy?" I asked. "Can you put it down on the house?"

He glanced at me, studied the tree, looked at the house, and nodded. "Yeah, I guess."

He worked his chainsaw like an experienced surgeon works a scalpel. Cut a wedge here and another one there, go deep, back off, nip again from the other side. The pine listed toward the roof's center line, then lurched with a groan and hung, swaying in mid-air. One last cut.

Snap. Whoosh. Pooofff. The pine crashed onto the roof; the center post gave way; the whole structure slid sideways; dust, timber, and old insulation exploded outward. With its branches waving gently, the pine came to rest a few feet off the ground, nestling the roof like a mother hen protecting her chicks.

Eddy stood near the remaining stump with the growling, puttering saw still in his hand. He nodded.

"Nice cutting," I said.

"Yeah, it came down to where I wanted," he replied with a sly smile. "I thought it would."

After watching Eddy drop many trees, I've learned to work a chainsaw fairly well. Last fall, I cut down some big oaks to restock my woodshed. They all fell right where I wanted.

Taking a Walk

One afternoon, in the process of recovering from colon surgery, I resolved to take a walk through our woods. I'd been out of the hospital for less than a week and I was still hurting. I'm a grouchy patient, in large measure because I'm impatient with the limits imposed by infirmity. I've never had to deal with chronic pain or a long-lasting illness of my own; in that way, I've been lucky. This would be my first long walk after arriving home and I wasn't in the mood for company. But, as I left the house, Eddy unexpectedly arrived.

"I'm out of sorts and taking a walk. Priscilla's in town," I said, as he rolled down his truck window. I hoped he'd get the hint and head back home.

"Okay," he replied, turning off the truck and getting out. He was going to join me. I looked away, resigned to my fate.

We set off. He could see I was in some amount of pain: I walked slowly, pausing several times on the steep hill behind the house. Where other folk might have offered to help in some way or asked me how I was doing or chattered away to cover their own discomfort with my feebleness, Eddy didn't say anything. He just walked when I walked and stood patiently when

I paused. We finally reached the woods and found a couple of fallen trees to sit on. We'd exchanged only a few words.

"It's a beautiful day," I said, lowering myself gingerly. The sky seemed particularly blue and the air especially fragrant with cedar scent. The land seemed to welcome me as much as I welcomed it. My pain seeped away just a bit.

"Yes, it is," he replied.

We sat in comfortable silence for quite a while. And then I started to tell him about the surgery and how it reminded me of life's fragility; I rattled on about my worries for the future of the farm and its sustainability, and for my children and their futures. I spoke of small irritating events in my life and major existential threats to our world.

Eddy sat still, as if he were waiting for deer or squirrel to wander by. He seemed to be listening. At one point he said, "Oh, it's not that bad." But otherwise, he remained silent. He didn't try to make me feel better. He didn't try to cover my gloom with a veneer of positivity. His quiet presence was a gift I didn't know I needed.

After a while, a sharp stab of pain reminded me that healing takes time, that I would have to put up with infirmity for a while longer. I stood up slowly. Eddy followed, waiting to see if I needed help. We wandered back to the house and stopped next to his truck.

"Well," he said, "I'll get out of here." He slid behind the wheel and started the engine.

"Take care, Eddy. Thanks for stopping by."

"You take care, too, Henry."

No need to say more.

Thanksgiving Squirrel

On the Monday before Thanksgiving a few years ago, Eddy dumped the contents of a black plastic bag onto our kitchen countertop. Twelve frozen squirrel carcasses without heads and feet slid across the surface.

"Here's the squirrel."

"Well, Eddy," Priscilla replied, "These look good. I'd planned to make our usual stew, but maybe we should do something different this year. I could fry them instead, except I've never fried squirrel. Plenty of catfish and chicken, but not squirrel."

"Mom always fried 'em," he announced. "It ain't hard. Just batter them good first. Salt. Pepper. That's all they need. You got lard?"

"Yeah, I've got some fresh lard. Okay, I'll fry them up this year. You'll come over, right?"

Eddy usually came over for dinner the night before Thanksgiving, when Priscilla made squirrel or venison stew depending on what he'd decided to hunt. And most years, he also sat down with us for our Thanksgiving meal, joining a collection of our friends, children, and children's friends. He became a familiar part of our celebration of community. Like all of us, he'd wear his best outfit: a flannel shirt and clean jeans. He'd put on his best shoes, shave clean, and comb his hair.

One summer afternoon, I was standing in our front yard with a visitor, a young lawyer from Washington, DC, who'd come to the farm with one of my colleagues from work. Eddy drove up (unexpectedly) after a day's work sawing trees for a lumber company. He hadn't cut his hair, or shaved, or changed his clothes for a few days. He looked as fierce, hard eyed, and wild as a hunting coyote.

Eddy parked his truck, climbed out, and walked slowly over to where we were standing. He nodded at me and I introduced him to my visitor. Eddy tilted his head slightly in the visitor's direction.

"He went and sold his land," Eddy announced.

"You mean our neighbor?" I replied.

"Yeah, the new owner, he's from Winchester. I done stopped to talk to him just now. Probably a developer."

My visitor struggled to understand who Eddy was. An uneducated mountain man? Maybe a West Virginia hillbilly (whatever that meant)? A reckless, gun-loving redneck? All of these things? Later, I gathered that our visitor, who'd lived in big cities all his life, had never met someone quite as rough-hewn as Eddy. He realized eventually that Eddy was part of the farm, he was from the local community, he was of these mountains. I could have been my visitor once upon a time, but now I've lived here too long. In a small way, I'm part of the local community now. I'm not from these mountains, but they're in me nonetheless.

For many reasons, Eddy could have decided to be bitter about his life or quick to judge others or always ready for a fight. At some point in his life, maybe he was those things, but he's not now. He'd say his life has been tough but good. He'd say he hasn't had a lot of opportunities. Still, he's lived a life mostly of his own choosing and on his own terms. For better or worse, he won't work for anyone he doesn't like.

I've watched him deal with people who aren't from the mountains. If they brush him off or pretend he's not there, he's liable to say something to me later, something like: "Oh, they're just city folk who probably don't know how we do things out here. They probably never did eat squirrel."

Friendship, After All

I tell about Eddy because being around him—working on the farm, timbering in our woods, or hunting with him—offers me another way to understand this land. His stories about his family, the wild animals he's encountered, the seasons he's spent walking the mountain all roll into a single narrative, a winding tale that's led him to a quiet patience with the world's contradictions and convulsions. Seeing our place through his eyes reminds me that humans can live in practical harmony with the natural world, neither disrespecting it nor making it precious.

Over the years, my relationship with Eddy has deepened the mystery of friendship. Ours is an unusual one—for him as much as for me—because our backgrounds couldn't be more dissimilar: where we grew up (New York City/Paw Paw, West Virginia), where we worked for a living (at a desk/outside), where we spent free time (city streets/mountain trails). But for more than two decades, my friendship with Eddy has endured. He still helps me put the bush hog on the tractor. I still help him renew his health insurance.

Perhaps all true friendships are one of a kind. The few I have differ in their mix of familiarity, affection, and the needs we meet for each other. They're all important in their own particular way. Each one bridges a space between different views of the world, even when that space is sometimes a canyon. Each one has thrown out a lifeline when I've gone overboard in rough seas, although that lifeline has sometimes been brittle. Each one offers companionship without the intractable complexity of romance. Call it luck or call it fate, my relationship with Eddy reminds me that some friendships arise and stay true for mysterious reasons.

I think Eddy would agree. As he might put it, "Yeah, that happens."

Early Years

Priscilla

Eddy is built like a wrestler, low to the ground with broad shoulders and hands that are surprisingly strong and flexible, even though they look as if they're carved out of rock. When he grabbed a goat, it stayed put. His instincts came from generations of mountain folk and served us well on the farm and in the natural world.

I started the goat farm with a dozen meat goats trucked in from Montana. Before they arrived, I asked Eddy if he would work for me. I needed someone strong, kind, and honest. Eddy was all of that.

I also had the aid of our guardian dog, Hercules, who was loyal and fearless and had no idea what I wanted from him. Just like Eddy. None of us had done anything like this, which didn't seem to bother us. We were a team on a new adventure.

Eddy had been around farm animals his whole life. That was the good news and the bad news. At first, his methods of handling animals were too rough for my management style. Eddy's way of catching a goat was to tackle it like a linebacker. Not gentle and sometimes not effective. Luckily, he was open to learning a new approach. As we worked together with the goats and dogs, he realized that a gentler, more patient hand was easier on all of us. He learned quickly and we became a good farm crew.

We had a hard time in the beginning until I realized he was deaf on his left side. He never would have told me. He just figured it was the way things were and none of my business. Or maybe he didn't even realize it. Knowing the rural culture of my area, I never asked why he had hearing issues. That would've been poor manners. Especially coming from a woman.

He'd fired shotguns since he was small without a thought about protecting his ears. He'd also logged with his dad, using loud chainsaws. After many friends and family fussed at him, he finally started wearing protective head gear.

Once I understood the situation, I tried to position myself on his right side when I needed to talk to him. I still had to yell, but most of the time I got a reaction. We muddled through.

Patience is such an important part of handling animals. The slower, gentler management brought out some interesting aspects in Eddy's character: he started to anticipate some of the goats' actions, he didn't always have to rush to the next thing, he was more comfortable taking time to make his decisions—and most of his decisions were right. After a while, I could leave him with the herd. He easily won the dogs' trust and affection. He became a true shepherd, and all the animals trusted him.

Eddy grew up with guns, mostly for hunting. He told stories about how he and his brothers brought home squirrels for stew—not just for sport, but because they needed supper. His skill with a gun came in handy for life on the farm. When we had to put an animal down or kill a dangerous pest or predator, he was fast, accurate, and kind.

Finally, we got to the end of our first year. I culled my herd, selling some of the goats and keeping my favorites. We also had two puny wethers I wanted to have butchered. Part of my plan for the farm involved developing a small goat-meat business, so I needed to learn how to cook it right—and eat it—to understand what I was selling.

I made some calls to professional butchers within fifty miles and discovered that most of them did not want to handle goats. Since I had few choices, and they all seemed about the same as far as cost, I chose the closest one.

Eddy and I loaded the two goats into the truck. He was very quiet as we drove to the butcher. I asked if he wanted to stop for ice cream on the way back to the farm. All he said was, "Sure, I guess." Something was up with him.

"Eddy, you have something on your mind. Would you like to talk about it?" There was a long silence between us, and I made myself stay quiet.

Finally, he said, "I know we talked about the meat business, but I never thought about butchering these two we're taking. They've been like pets, and they're healthy. It's not like they done something wrong."

"Eddy, you know I can't sell these two for anything but meat. I need to learn about goat meat." I rolled my eyes. "I'm not running a petting zoo." I was sorry about the exasperated tone in my voice and the sharpness of my words, but we had talked about this. He never questioned our plan before.

He stared out the window the rest of the way. We didn't talk.

We found the butcher without a problem and I parked the truck and walked into the front office. Inside were a messy desk and a display cooler in the corner with only a few cuts of meat for sale. The place looked like an unkempt, down-on-its-luck business. A scruffy-looking guy came out of a door, shoved a cut sheet in my hand, and pointed to the messy desk and chair. I filled out the form and left it on the desk. The man stood in the corner staring into space.

"Can you tell me the kill day for my animals?" I asked.

"No," he replied, barely looking at me. "I'll have the boss call you. Drive around to the back and put 'em in any stall."

Eddy and I drove the truck to the back of the building and looked over the stalls. Even though the barn was dark, we could see it was filthy and empty. Many of the stalls were heaped with manure. We searched for water buckets and a hose, but didn't find any.

I finally decided on a stall, the cleanest dirty one. We brought the two goats from the trailer. As we closed the stall gate, its top hinge fell off. *Not a good sign.* Eddy managed to bang it back on.

We were in the barn long enough to be sickened by the stench of filthy livestock holding pens.

We walked to the truck in silence and started home. We were almost out of the driveway when Eddy said, "Stop."

I braked and looked at Eddy, puzzled by his strong reaction. His blue eyes glistened with tears; his voice was shaky but clear. "We worked hard to treat those goats right and gentle," he said. "I can't leave 'em in that filthy place. They deserve better."

Without saying a word, I turned the truck around and pulled up to the barn. Eddy marched straight to the barn to rescue the goats while I walked back into the office and told the man we were taking our animals. I also told him his place was terrible, and he should be ashamed.

As we drove away with the goats in the trailer again, Eddy had an idea. His cousin butchered deer for extra money in the fall. Why not see if he'd handle the goats? They have similar bodies and skeletons. Maybe he'd do us a favor.

When we arrived at his cousin's house, it was modern and clean. He was a nice guy and happy to help.

He and Eddy took the goats behind the building. Eddy shot them right there. One bullet. No suffering. The two men worked together to dress and hang the carcasses.

As I waited for Eddy and his cousin to finish, I wondered about that first place. If I had been alone, would I have gone back for the goats? Would I have left them?

As we drove home, Eddy spoke up. He believed in our work, believed the animals he helped raise were important enough to take a stand. He believed our livestock were worth respect throughout life and in death.

The next year we qualified to become a member of Animal Welfare Approved (AWA), an international organization committed to the humane treatment of farm animals. Every six months, AWA audited our farm to assess compliance to their rules. We were all proud when we kept getting prefect scores for our farm management.

Eddy had come to understand the importance of humane treatment of animals and the environment. He walked and loved the mountains surrounding us. He gave me a better understanding of the culture around me. We both realized the job we were doing was a balancing act; sometimes you got it right and sometimes you didn't. Farming is full of trying and not making it, getting up, and trying again. But we had clear standards. They were our map.

Eddy remained my left-hand man for many years, and my trust in him and his opinions only grew as time went by.

Hercules

Priscilla

I drove down the steep driveway. It wasn't very long, but it was narrow and bumpy. It ended about twenty feet from a tiny, rickety pen that held a big puppy and a wiry-looking buck.

The pup stood his ground and barked in a low tone as my truck came into his territory. He showed no fear, just a steadfast resolve to do what he had been bred to do.

That was my first look at Hercules, a three-month-old male Maremma pup. The thick snow-white fur that covered his body stood straight up on the top of his handsome square head and down his back. His black nose was like a piece of coal in the middle of a winter field. He had almond-shaped hazel eyes. He was only a pup, but his instincts were clear to me. He would do what needed to be done to protect whatever was in his care, even if his side still hurt from the beating the buck gave him after the farmer put him in the pen.

I climbed from my truck and looked at the young man who owned the place. "Why is he in a pen with a buck?"

The man shrugged. "Because he chased the kids and the does." The doe herd and kids were in another pen that was slightly better than the miniscule one the pup and the buck were in.

The young couple who owned the place had no idea what they were doing. They were barely out of their teens with two children under three. The wife was pregnant again with their third child.

A menagerie of animals ran everywhere, with the little boys chasing them, waving sticks over their heads. The mother looked on with no emotion, just a lot of fatigue. I figured they had maybe eight acres. The trailer they lived in was small for a family with one child, but coming on three was unimaginable.

He told me he was going to be a farmer and live off the land. "Bad coyote problem around here," he said. He was told he needed a guardian dog to protect his goats. "I bought this pup from an older couple up the road. They said the dogs were guardian dogs, called Maremma. Come from Italy." He was proud of his knowledge.

I asked what he wanted for the pup. "He ain't done nothing, and I paid a hundred dollars for him!" He stared me in the face, ready for an argument. "That's what I'm asking for him."

"I'll take him," I said. "Would you open the gate?"

"Oh, lady, he'll run off. It happened the other day. I had a terrible time getting him back."

"Open the gate," I repeated. "I'll take responsibility." I sat down on the ground, crossed my legs, and hoped for what I figured would be a miracle.

"What are you doing?" the young man asked.

"I want to see if he will come to me," I answered.

The young man shook his head and slowly opened the gate.

The pup cautiously came out of the pen. He stood still for a bit and looked around. After studying the area, he walked up to me, curled himself up, and lay down in my lap. His big body trembled. I wrapped my arms around his back. He continued to shake but didn't move from my lap.

Despite his weight, I rose to my feet, keeping the huge dog in my arms. I put him on the backseat of the truck. He lay across the seat as if he'd done it a hundred times. I wrote the young man a check and took Hercules home.

He worked out well in my family, which included Sammy, a big lab-shepherd-mix dog, and two teenaged boys. For the next eight and a half years, Hercules had my back. He was my four-legged knight in white fur.

When we first bought the West Virginia farm, we lived in a rural county north of Baltimore. Because of the boys' school and my work, we had to be close to the city. That's also when Henry began staying in DC during the week for his work. Hercules figured he needed to patrol not only the horses but also the people. His family was not intact with Henry gone. He would walk the fence line of our three acres at night and howl, a trait used by guardian dogs to let predators know they were on duty.

We had some farmers around us, and mainly they understood what he was doing. Most of the neighbors were thankful. Some were not too happy. But we never had a problem with predators in the neighborhood.

Hercules worked on that same goal when we came to live on the farm full-time. He prowled the fields of the farm at night, sometimes howling as he went. We did have a complaint from the woman on the next farm over. I never worried, though; her husband, the farmer, was pleased not to worry about predators. We didn't have a coyote problem that winter—and neither did our neighbors. We didn't lose a goat, and my neighbor never lost a calf. We also never got another complaint about nocturnal barking.

Early spring brought our first kidding season on the farm. As soon as the first kid hit the ground, Hercules knew what to do. One of his most important jobs was serving as bookkeeper of the new kids born every year. He thoroughly examined each one, from sticking his nose up their butts to licking their heads—despite the battering and bites he received from their mothers. Smelling the babies gave him information that went into the database in his brain. He was the guardian of that kid for as long as he or she was on the farm.

By their second season of births, most of the does let Hercules go through his ritual. First sniff, then lick. Their kids seemed none the worse for it. The routine went on until all newborns were inventoried. Hercules never left the does and kids after the first kidding season, rotating with them. He would go from one group to another, checking on his goats throughout the day and night.

The herd of does quickly learned babysitting was one of the perks of having Hercules around. He always picked nice, dry places for naps, and the kids and he all slept wrapped around each other in the shade. He never moved a muscle until all the kids had stood up, stretched, and run to play or find mom for a snack.

Before we had the goats, he had loved to come inside, but now Hercules found a spot in the pasture above the house where he could see the barns, the fields, and the house with just a turn of his big head. He slept there for years, staying on watch. When the boys came home from college, they would try to coax him down from the hill and into the house.

"Mom, why won't he come in? He'll freeze in that snow,"

they'd exclaim. Once again, I defended Hercules and my methods of farm management.

My older son would yell, "One morning we're going to find him frozen! Have you thought of that, Mom?"

"He knows what he's doing," I'd yell back.

The next morning, they would always see him in the pasture weaving among the herd like a huge white ghost. Not one goat paid him any mind as he silently passed through the snow-covered pastures on his early morning rounds.

Late one summer, I moved the livestock to the pond pasture where the pine thicket added a cool break from the heat. I was filling mineral buckets for the goats when Hercules stood up, put his nose in the air, and walked slowly, silently, to the end of the pasture. The hair on his back stuck straight up.

He scrutinized the woods near the creek—never making a sound, just staring through the fence. The undergrowth on the other side of the fence was thick. The woods behind the thicket butted up to the mountain that bordered the back of the farm. There was a gentle breeze, not strong enough to make much noise. I was puzzled and couldn't see what had his attention. He moved closer to the fence and stared through the underbrush into woods.

After a few minutes, a strong, low howl came from deep in the woods, followed by a long silence.

Hercules never moved but continued to look in the direction of the howl.

After another minute or so, there came another howl. This time, Hercules answered with his own call. He continued to sit and stare in the same direction.

By this time, the goats and I knew something was up. Everyone stopped and looked toward Hercules. The silence was unnerving.

Adrenalin rushed through my body. Were we in danger? I had never heard coyotes that close, and I was glad I had the truck nearby. But what about the does and the kids? There was only one Hercules, and I didn't know how many coyotes were roaming the mountain. Their noisy yipping was spooky as they romped down the mountain with their pups.

Hercules went into a defensive trot around his herd and the perimeter of the field. Now and then he would let out his own howl, followed by barks, and wait for a reply. Was he trying to judge the location of the pack? Or was he sending a message to the leader of the pack? Either way, things were tense.

Another long, mournful howl came from the mountain. Hercules stood still. Pointing his head at the sky, he let out a long, strong howl that ended in a menacing low growl. His eyes narrowed and his head dropped square with his shoulders. His body language changed to that of a warrior ready to defend his herd. The sounds coming out of him were primal and sobering. He was a large, fierce animal that would fight to the death, if necessary.

Then silence.

After a few moments, the pack moved off. High-pitched whines and barks came from the females and the pups. The voices carried up onto the mountain. We stayed quiet and still, until their sounds faded into the hills. My legs felt like jelly. I realized I was clutching my shovel like a weapon; my hands hurt from the pressure.

A bargain of some sort had been reached between Hercules and the pack leader because we didn't hear the big male again until winter set in. The pack made sure we knew they were in the mountains, but they stayed out of reach, and that was fine.

After that, I decided Hercules needed help with a herd the size of ours. The farm backed up to a mountain range that was still wild and used only for hunting. We needed another guardian dog.

Our next Maremma was Bella, a fuzzy white pup that was born in a barn full of sheep. I brought her home at eight weeks old. Hercules sniffed her all over and decided this was not help, but another baby to look after. He looked at me quizzically.

"It will work out," I assured him. "She'll grow up and help you." And, eventually, she did.

Bella had been with sheep from birth, so that helped with her training. She took to sleeping with the herd quickly and wasn't interested in life outside of the herd. But puppies are puppies and they need toys, and a goat's wagging tail looked a lot like a toy. Thus, unfortunately, some goats had sore tails during Bella's baby years. Hercules had his hands full teaching her, but by the time she was a little over a year old, they had become quite the dynamic duo. She was no longer a cute little pup that curled up with her goats—she was an important part of the team.

Most guardian dogs pick favorite kids now and then, and Hercules had a kid he fell in love with. It was a little doeling born to a wonderful and patient Spanish doe named Betty. We named the doeling Pearl for her markings and coloring. She was feminine with long strong legs and had a rare striping on her face. Hercules was completely devoted to Pearl, who followed him as much as she did her mother. She was a healthy young kid who seemed to enjoy life. If she hadn't slept safely curled up in his warm fur through the night, they found each other first thing come morning.

At three months, kids should graze with the herd, but Pearl shied away from the chaos. We fed her and Betty separately, but she still didn't seem interested in the hay, grain, or the pasture. We couldn't figure out why she wasn't thriving. We wormed her and gave her vitamins, and I even called the vet.

Pearl would rally, but then she'd get weak again and not eat except for nursing from her mom. Most does don't let their kids nurse after a certain age, but Betty let Pearl nurse whenever she needed. After a while, Pearl was too weak to follow the herd in the pastures and wouldn't leave the barn. Her mom and Hercules stayed with her, taking turns going out, Betty to graze and Hercules to help Bella keep track of the herd.

After a few days of this, Betty left Pearl in the barn. She knew Pearl was dying. It was also fall and Betty felt the drive from nature to breed with the buck that had been put out with her herd. But Hercules stayed. When he left the barn to check on how Bella and the herd were doing, he always rushed back to Pearl as soon as he could. It got so we took Hercules's food to the barn where Pearl lay. He doted on her night and day.

Henry and I decided to let nature take its course. We would no longer doctor Pearl. She couldn't live long. I think even Hercules knew. Nevertheless, he would not leave her and became aggressive even to the barn cats as they strolled by. On the last day of her life, he stayed with her every minute.

I had to get on with chores but tried to stay within earshot of the goat barn. Around four o'clock in the afternoon, a mournful howl came from the barn, followed by whining and frenzied barking. I knew Pearl was dead. I also knew Hercules would be hard pressed to give up Pearl's body or let anyone near her for a while. He needed to deal with his sorrow, so I left things alone. He didn't leave Pearl's body for two days. He nudged her body gently, laid his head on her, and whined.

Finally, he left the barn, and I was able to take Pearl's body. I called the vet, who had always been puzzled by Pearl's illness, and he ran several tests. We found out that she had a birth defect in her colon, which was twisted, and as she grew and needed to eat solid foods, it became a real problem. The food couldn't get by the blockage and pass through her digestive system. That's why she could drink water and mother's milk, but that was about it.

Hercules went back to his job, but he never took up with another kid like he did Pearl. He slept alone in the pasture overlooking the house and barn.

It was spring and the kids were starting to be born. This was my and Hercules's favorite time on the farm. He was a little past eight years old and in his prime. One day I noticed a knot on the side of Hercules's leg. Nothing else. He seemed normal, eating, and active as ever, so we went about our lives. I decided not to worry, after all he was in his prime: strong, smart, and amazingly beautiful.

When the dogs' yearly shots came due, I pointed out the knot to the vet. He suggested we do a biopsy. It came back as cancer. There was nothing we could do.

Hercules went about his chores as actively as he could, and so did I. He was getting tired and sleeping more. The other dogs let him have his naps and took care of the herd without him, and we all put up with his bad humor. He was short with the kids that tried to sleep on and around him. That had always been one of his most beloved jobs. His food was rarely finished and when the other animals came by to eat some, he didn't seem to care.

I knew the dreaded morning would come when I would see the weakness and fatigue in Hercules's eyes. When it came, he didn't want to eat and was mad at the world. He slept on the porch of the house during the day which he had not done since he was a pup. We would find him in a pasture sound asleep, not knowing the herd had moved on to another spot on the farm. His moaning sounded painful and sad. He stayed with the herd trying to do his job with as much energy as he could muster. He still climbed the hill where he had slept for years, but every night the climb grew harder and some nights he didn't make the top. As much as I hated it, I knew it was time. I called the vet.

One of the magical things that happened that terrible day was that both of our sons were home. My fantasy was that Hercules called them to come home. "You and Dad will need them," Hercules would say. I also knew he wanted to see his boys one more time, even though he couldn't run down the hill to them as he always had, swinging his fluffy tail like a propeller.

As he watched them walk up the hill, he wagged his tail and sat up with as much dignity and strength as he could manage. They sat on the ground beside him, gave him their last hugs, and just hung out together.

When the vet got to the farm, he and Henry walked deliberately up to the goat barn where Hercules was lying in a small grassy spot. Even Sammy, our ancient lab-shepherd mix, came with them. Bella lay beside Hercules, licking his face. I looked out into the big pasture where the herd of goats and the other dogs lay on the grass in solemn gratitude: they had come to pay their respects.

It was crazy how beautiful an early spring day it was. A glorious blue sky, gentle breeze, and the silence one only gets from the quiet of the country. Hercules's family was all there, the boys, Henry and even Eddy. Bella and I were right beside him.

Hercules never moved as I sat beside him, stroking his big noble face. He looked at me with tired, sad eyes, and told me goodbye. The vet gave him the injection. He let out a big sigh and was gone.

None of the animals moved, except for Bella, who licked Hercules's face and whined her sorrow. There wasn't any other sound. The king was gone, and everyone knew it. I could only lay my head on his majestic body and weep. I had started all this with him. How did I keep my dream alive without him? His was a loss I had never planned for. Who was going to have my back now?

After a while, the herd and the dogs quietly moved farther out into the pasture. Henry and the boys rounded up what we needed for Hercules's burial. The vet and Eddy left. Only Bella and I remained beside Hercules. What was my next step? How did I fill his paws? Had I learned enough to go on without him? I couldn't take my eyes off him—I had so little time left to gaze at him, and then he would be gone for good. Bella slowly got up and went into the field with the herd and the other dogs.

Henry drove the tractor over, and the boys helped put Hercules in the bucket. I wrapped him in my favorite hand painted velvet quilt I had made years before. We buried him on the hill where he had slept and watched over all of us for so long. Bella lay on his grave that night. When her howl went out in the dark, I knew she was saying goodbye to her love.

A month later our oldest son, Nick made a grave maker that still stands where Hercules is buried.

Bella was in the field with her herd the next morning. She moved silently through the goats like we had seen Hercules do for years. Bella was dog boss now. At times, I was sure she wondered what Hercules would do about an issue she faced.

Farming is mainly chores, chores, and hope—but then there are moments of memorable magic like a four-legged knight in white fur.

Morels on the Mountain

Henry

The spring of 2022 brought relentless images of despair: Ukraine's grandmothers, with deeply wrinkled faces, weeping as they stood beside dead bodies; grieving parents clutched to each other, pleading for solutions to school shootings; young children at various border crossings, looking for hope with soft, bewildered eyes. So on an April morning, I left this world's turmoil to search for morels for a time, and, in my searching, stumbled upon an outcropping of grace.

Morels are elusive mushrooms. Tinted brown like dried leaves and squatting close to the ground with pitted caps, they're easy to miss. They also rarely show up in the same place from one year to the next. But, as I saddled up my trail horse, I had reason to hope I'd find enough for lunch. I had a clue to their whereabouts from Eddy, our friend and neighbor, who's walked the surrounding hills and hollows all his life. He knows the mountain's shoulders, sinews, and veins as well as he knows the contours of his coarsened hands. Eddy had gone morel hunting with great success the previous day and had stopped by to show us his treasures—one as massive as his fist. A generally sober mountain man of few words, he was uncharacteristically excited.

"Never seen one bigger," he said. "And I've been finding 'em for fifty years. There's a lot of them about this year. Ain't it pretty?"

And it was beautiful. Its beige veins wove a riotous maze of small alleyways dead-ending in shallow pits. Magnificent and mysterious, its convolutions were a cross between the surface of a sponge and the folds of a brain.

"Where'd you find it?" I asked as we stood together in my front yard.

"Up yonder," he replied, gesturing toward the mountain behind the farm.

"Yeah, Eddy, I know you found it on the mountain. But, like, maybe you could be more specific. If there's a lot this year, I'm not going to find all of them. And I won't tell anyone else."

Although we've known each other for twenty years, spending time together building barns and clearing fields, he glanced at me like I was a big-city lawyer just arrived to fleece him from his land. A long pause. He wasn't going to make it easy.

"Eddy, look, you can come over for dinner if I get enough. Steak, mashed potatoes, and fresh sautéed morels."

A few more moments passed before he shrugged and muttered, "Near the run bed close to Boozer's cabin."

That stream stretched halfway up the mountain, so it wasn't much of a hint. But I wasn't going to get any more. It was enough. I knew where to start looking.

I saddled my horse, Gabe, and checked for carrots, knife, and mushroom bag. We headed for the old forest road behind the back pasture. About twenty minutes later, we'd climbed to where the road, now overgrown and blocked by the occasional fallen tree, intersected the stream that Eddy had mentioned. I dismounted, tied Gabe to a tree, and followed the stream downhill. Hunched over, I walked slowly, pushing the brown, papery leaves aside with my boot. Where the stream had run over its shallow bank, I bent down to flip large pancakes of wet leaves. No morels. I checked around all the dead trees and at the base of every nearby oak, elm, and ash. No morels.

After a while, Gabe snorted. I looked up. He was pawing the ground. He'd eaten everything he could reach and was ready to move on.

"Just stand there," I hollered and climbed back up, retracing my steps along the stream and looking again at every likely spot.

I moved Gabe to a tree near another patch of grass, fed him a carrot, and told him he'd have to be patient. I headed uphill. Twisting tightly among the trees above the old road, the run bed was steep and rocky. I crawled along, brushing aside leaves and small limbs with my hands and bruising my knees on the unforgiving rocks. No morels.

Was I that oblivious? Was Eddy so paranoid he'd sent me in the wrong direction? Were they just not there? The perfume of damp earth and decaying leaves mixed with saccharine scents of autumn olive. I was hot and irritated.

Gabe snorted again, ready to move on. Frustrated and unwilling to return empty-handed, I untied Gabe, swung myself up, and followed the road as it climbed upward. I'd been there before, a long time ago, but couldn't recall where the road came out on the ridgetop. Gabe must have remembered because he never hesitated, slipping past blackberry bushes and hopping over fallen trees. The trail vanished into a rock fall; Gabe picked his way through.

Finally, at the ridgetop, the trail leveled out in a pine grove near a rock ledge. I dismounted again and we walked forward toward the daylight showing through the trees—rather than the expected forest shade. To my surprise, the mountainside below the ridge had been clear cut. Virtually every tree was gone.

I didn't know the owners, but they must have needed the timber money. Maybe they were cashing in before selling the land. Maybe houses would be here the next time I came by. Still, why couldn't they have left the younger hardwoods?

I led Gabe through a few windblown trees, dodging some jagged rocks. We stepped over the exposed roots of the last pine, and I raised my eyes. There it was: The View. The timbering had felled the trees that would have otherwise obscured the space in front of me. I gazed into the vast blueness above the corrugated surface of Appalachia—mountains undulating into mountains over and over until the last ripple faded into the horizon's fuzzy gray. Shadows of a few marshmallow clouds drifted across the nearest hillside. Far below, the meandering Potomac sparkled with fleeting glints of sunlight. I seemed to be standing in the narthex of an immense cathedral.

Taken utterly by the enormity of the sky and stunned by the distant mountain's stateliness, I drifted into a deep reverence. Gabe and I seemed tiny and insignificant. A stiff breeze, rustling the pines behind me, swept away my self-absorbed peevishness and left an age-old question: How can the breathtaking beauty of this world coexist with its horrifying brutality?

As much as they inspire a reverential awe, grand vistas also discomfort me because they prompt questions I want to answer but can't. What would happen if the leaders of every country spent time every week pondering this view? Why is it that these views enlarge our capacity for wonder? What would happen if every young adult had to serve two years on such an overlook as guardians of the natural world? And perhaps the most urgent question: How am I to respond when nothing I do seems to lessen the violence—not only between cultures, but also between the human and the nonhuman? No religion has yielded answers that ring true to me or that quench the smoldering agony of impotence. I live with the mystery that lays behind these questions.

Judith Thurman, in a *New Yorker* review of translations of Dante's *Divine Comedy*, made this observation: "His medieval theology isn't much consolation to a modern nonbeliever, yet his art and its truths feel more necessary than ever: that greater love for others is an antidote to the world's barbarism, that evil may be understood as a sin against love, and that a soul can't hope to dispel its anguish without first plumbing it."

As I stood on that ledge, a diminutive witness to the vast expanse of the natural world, I knew I wasn't going to change the news or alter the dismal mess we've made of our environment. But, somehow, the view shifted something inside. A bit of hope slid out from the blanket of despair, a hope that Dante was right: that loving deeply can mitigate the consequences of brutal trauma. Maybe small acts of kindness—gestures of genuine love—are sufficient responses to the evils of this world.

Lifting his head from scrubby grass, Gabe watched a hawk circling slowly below us over the shaved hillside, its stumps protruding like so much stubble on a giant's cheek. I saw Gabe stretch his neck up a bit, as if the distant mountains had just caught his attention. He didn't seem particularly impressed with the view. He shifted his weight and nudged my shoulder. I found a carrot, and he was happy.

Whoever says that venomous conflict is part of the human condition suffers from an impoverished imagination. The past does not have to be the future. Evil is not inevitable if we can wield the most powerful tool we have: our capacity to conceive a more compassionate world.

Gabe whinnied, as if to suggest that I'd plumbed my anguish enough. I was reluctant to leave the grace that surrounded me, but it was time to go. Obligations were whispering their demands. I'd have to live with all these questions. Maybe remembering this moment—writing about it—would suggest some answers.

I swung myself onto Gabe. We stood on the ledge for a few more moments. The breeze had vanished and the sun's heat was making him sweat. At least I could resolve to bear witness to the world's horror—to not look away from the despair. If I'm to appreciate the stunningly beautiful, I also must embrace the discomforting ugly.

We found our usual path down the mountain. An easy ride, clear of rocks and fallen trees. We crossed a stream on the opposite side of the cabin from where I'd previously looked for morels. I couldn't resist searching for a few more minutes, much to Gabe's disappointment. Right away, I saw a small one. Then another. And a third. I kept searching until Gabe snorted hard, but I found no more.

Back at the house, I showed the morels to Priscilla. She was unimpressed.

"All that time and only three puny ones." She shook her head. "Oh, well," she added with a sigh, "They'll be enough for lunch." She served the sautéed morels on a small plate. Shriveled from the heat, they looked pitiful. We each had a few bites, relishing their nutty, almost smoky, taste. I told her about the clear cut and the view.

"You've been communing with the land again," she said, knowing my habits well. "Mushroom picking and mountain gazing."

"Yeah, I guess so."

"Did it help?"

I paused before answering her . . . "Just enough."

Arnost and the Eagle

Priscilla

My goal was to shell the beautiful mess of peas I'd picked yesterday, bag them, and store them in the freezer. Not exactly what I wanted to do, but I'd raised a good crop, and I'd be glad to have them come winter.

But I couldn't get my mind off the injured kid in the pasture. He was Freda's first born. I wasn't sure how the kid had sprained his leg so badly. I thought it was during one of the death-defying leaps the kids take off the big rocks piled near the goat barn. I needed to check on him. My worry grew and the lovely morning air beckoned. The peas would have to wait. I promised myself I'd finish them when I got back.

As I walked to the pasture, I passed Arnost, one of my guardian dogs, dozing in the morning sun. Even for a Karakachan, he was big. He had dark brindle fur that gave him a menacing tough-guy look. He had everyone fooled with his looks, until they saw how he was with the kids and grown goats. He was one of our gentle giants; gentle until they needed to be different. I learned when he was young that he was all about business.

I smiled as he moaned and twitched in his sleep. Behind him, the does grazed while their kids mostly slept. They all looked fine. But I still didn't see the kid I was looking for, or his mother. I walked through the pasture gate and checked one pile of napping baby goats after another.

Making my way up the hill, I noticed several does jerk their heads up and look toward something behind me. A savage growl interrupted the quiet morning. The hairs on my arms stood up as I swung around. Arnost was fully awake and looking skyward. In one graceful movement, he leapt up and ran hell-bent past me into the field. His growl grew into a warning howl as he followed something in the sky.

I shaded my eyes so I could see what had made Arnost so upset. There they were: eagles circling slowly.

"No!" I yelled.

The eagles must have seen the baby goats sleeping all over the pasture. An easy breakfast was on their minds. I ran from one pile of sleeping kids to the next, waking them, yelling as loud as I could, and waving my arms frantically to alert the does.

As the herd realized the danger, the scene became chaotic. Our two other guardian dogs ran in circles, barking and snarling at the sky. Does bleated for their babies; kids screamed for their moms.

In a few moments, most of the herd bolted past me, rushing for the safety of the barn. The dogs kept jumping and snarling at the slowly circling eagles, but Arnost stayed low: crouching, tracking them, moving to keep them in his sight, waiting to see if they dared dive for a kid.

I still didn't see my injured baby and her mom Freda, but from somewhere over the steep hill in the back pasture came more barking, deep and repeated. I ran toward the sound. As I reached the hilltop, I saw Sophie, another guardian dog, at the bottom of the hill, leaning over something that wasn't moving.

Her long, snow-white fur stood straight up along her back. Her big, strong body squared off on top of something on the ground. Her head darted up at the sky and she looked all around. As I got closer, I saw she was standing over the injured kid.

Freda was nearby—snorting, stomping her feet, and glaring at the eagles. She was there to defend her baby.

The eagles circled over Sophie until one of them dove right toward her. She crouched over the kid, growling and barking as the eagle descended. Sophie never moved from her post over the baby goat, even as the eagle came dangerously close. The eagle darted back into the sky to rethink its plan of attack.

At that moment, Arnost appeared at the crest of the hill, stopping to stare at the two eagles circling just above Sophie, the mother doe, and the injured kid.

Then, an eagle began another dive.

"Run, Arnost! Run!" I yelled, waving my arms as the eagle dropped toward Sophie and the goats.

"Get away!" I screamed. My heart throbbed as my feet flew down the hill.

The eagle plummeted, closing the distance to the kid.

Arnost was in a dead run, his strong legs pumping fast. He had only one chance. With all his might, he hurled himself into the air, going for the eagle. The bird saw him, but it was too late. Arnost grabbed the eagle in midair and brought him down. Together, the dog and eagle hit the ground with a thud.

I was still running with complete abandon, stumbling as I went, trying to watch the scene and keep my balance. My boots slid, and then I tumbled. Rock and dirt ground into my skin as my shirt tore and I rolled head over heels down the steep bank. I was out of control, frantic to stop my fall. I desperately reached out for a small tree and wrapped my arms around it as hard as I could. My shoulder snapped with pain, but I finally stopped rolling. As soon as I could, I sat up, spitting dirt out of my mouth, my face stinging from small cuts.

I heard the growls and screams of the two animals in battle. There was Sophie, still over the kid, dancing a protective dance and barking wildly. She hadn't moved an inch from the baby goat.

A few feet away, Arnost and the eagle were trying to kill each other. They were spinning in a blur of dust—screeching bird, snarling dog, thin blood ribbons on the ground and in the air.

Finally, the eagle freed itself from the dog's grip and jumped backward, opening a space between them. For a moment, it was a standoff, dog and eagle glaring at each other. And then the huge bird suddenly propelled itself straight up, jetting through the cloud of dust to the safety of the sky.

Feathers and blood covered the ground. Whose blood I couldn't tell. Arnost continued to growl savagely. The two eagles circled a few more times, drifted higher, and then disappeared. They would find breakfast somewhere else.

Sophie, still defending the hurt kid and her mother, sensed the fight was over and finally relaxed. She moved off the baby, licked her to make sure she was okay, and then nudged her up and over to Freda.

With the aid of a big stick, I worked to get on my feet. I hurt in so many places, but I decided to think about the pain later and looked around again.

The injured kid was between Freda and Sophie, so I staggered over to check them out first. All three animals were still shaking from the terror of the attack, but they were otherwise unharmed. I patted Sophie on the head and looked around the pasture. No other goats. The herd was all in the barn with the other dogs.

I steadied myself and thought about hobbling home. But where was Arnost?

As if he knew what I was thinking, he pushed his warm wet nose against my arm. He was right behind me. "Arnost! Are you okay, boy?"

I leaned on my stick, hugged his strong neck, and dipped my head into his thick, black fur. He felt sticky and smelt of blood and dirt. His legs were still trembling. As I held his big shoulders in my arms, I felt his heart racing, his mind still on the threat.

I picked a few feathers from his fur and noticed the blood on his face, neck, and front legs. Steadying myself with the stick, I leaned down and examined my war-weary Arnost. He had some nasty scratches, and I was sure he would be sore for a few days, but all in all he was okay. He looked back at me and sat in the grass, now able to relax.

"Man, we are a sight, Arnost!" My clothes were dirty and torn. Blood seeped from my cut. "Both of us covered in dirt, sweat, feathers, and blood," I said. "Just a sight."

Then my legs began to tremble, and my breath left. I dropped down beside Arnost. He and Sophie searched the sky for eagles. But they were gone, having learned their lesson—at least for a while.

I rubbed the eagle feathers in my hand as I pondered what could have happened—or what almost did. The guardian dogs did their job. That age-old method of protecting a herd worked.

I slid my hand along Arnost's fur as he lay in the grass licking his wounds.

After resting a few more minutes, I hobbled back to the house. My wounds needed some licking too. And, yes, I had to keep my promise to shell those blasted peas.

"Say What? No Way!"

Hog Gossip

Henry

One late August afternoon, I filled two large feed buckets for our hogs and marched to their pasture, resolved to consider them as simple farm animals rather than what they truly were: intelligent, fun-loving, and demanding creatures. I usually don't mind farm work, but that afternoon I was tired and a bit hungry myself—a condition that invariably fosters a simmering irritability. At that moment, the pig-feeding ritual seemed tedious and the buckets burdensome. I'd dump their dinner in the troughs, hurry back to the house, and check the refrigerator for leftovers.

Our four hogs—Iris, Ruby, White Foot, and Indie—had grown substantially larger since January, when we had bought them as piglets. Accordingly, the buckets of feed had grown heavier as well. Much heavier.

Nonetheless, I staggered up the hill and through the pasture gate, clanging the buckets against the gate's metal bars. The noise typically served as a dinner bell, prompting the hogs to sprint from their muddy cocktail lounge near the run-bed below. I poured the feed—a mix of grain, mashed potatoes, sausage gravy, and chicken bones—into their troughs and waited for them to jog over the hilltop.

Before moving to the farm, I'd never known a hog up close and personal. And so, after Priscilla found Tamworth piglets listed for sale in *The Market Bulletin* and decided we had to buy them, I educated myself about pig behavior. The only pertinent resources in my library at the time were two works by the American essayist E. B. White: his 1948 essay "Death of a Pig" and, naturally, *Charlotte's Web*. Neither provided much in the way of practical information, but they infused in me a spirit of generosity and forbearance toward the world of hogs.

Reading how-to books about hog farming helped, but watching four pigs actually grow up provided more tangible lessons. I hadn't realized how fastidious they are when given a chance to live in decent conditions. Our pigs divided their large pasture into specific areas: they mostly pooped in one place, relaxed in one of two muddy spots, foraged in the grassy areas, and ate dinner close to the gate. As long as you didn't stand between them and their food, they were friendly. And any observer could see our pigs relished human hands scratching their heads.

As I waited for the hogs to appear, I noticed one of the chicken legs still held a considerable chunk of meat. My stomach rumbled. Where were those self-absorbed slackers? They had never been this slow before. Ever.

And then an unsettling thought crept in.

What was that line from White's essay? Something like: "When a pig fails to appear at the trough for his supper, a chill wave of fear runs through the household." I waited a bit longer. No show.

Okay, something's wrong. Worry replaced my irritation.

I walked to the hilltop overlooking the run bed and scanned the scene below. There they were, gathered near the fence between their foraging area and the goats' pasture. They could easily have run up the hill for their dinner, as they had many times before. Instead, they were pacing around each other, quite agitated, but apparently not in imminent danger. A cluster of autumn olive bushes obscured the fence at that spot so I couldn't see what was occupying the pigs' attention.

Too tired to trudge all the way downhill, I whistled, figuring they'd just gallop up to their troughs with their usual boisterous anticipation. Not this time.

They glanced up and then looked back at each other. Indie turned and started to run in my direction. *Only Indie.* They had dispatched the messenger.

At first, I resisted anthropomorphizing our livestock—I tried not to attribute human characteristics to animals. That effort was especially fraught when it involved the hogs.

Part of the problem involved accounting for our pigs' specific "personalities." By mid-March, the four piglets had distinguished themselves as individuals, each with her or his own set of habits and mannerisms. The largest of the four, Iris, was the queen— and acted like it. She always ate first and claimed the choicest spot in the mud hole. Ruby was Iris' confidant; she deferred to Iris and liked to lie close by. White Foot had the occasional habit of turning precipitously in circles, her face expressing a sublime joy. Indie (short for Independence) was a male, neutered at birth and the smallest of the four; he generally kept to himself, probably to avoid being bossed around. More than any of the others, he loved having his head scratched.

Knowing them as distinct individuals, I could not dismiss them as just "hogs." I wasn't willing to lump them into a single, undifferentiated group. They were as different from each other as humans are. And they weren't just objects; they were agents in their own world, which was my world, too.

Language also made it hard not to anthropomorphize. I didn't have anything but human words to describe our livestock's behavior. I couldn't speak hog, goat, or horse. Still can't. But anyone who lives with livestock (or maybe most animals) soon becomes aware that animals of the same species communicate with each other, and, when necessary, with members of other species—including humans. Depending on the situation, they speak about what they are doing or feeling or thinking—or whatever words they'd use to describe those things. And they speak not just with utterances but gestures as well, moving their ears and faces and bodies in the service of messaging.

If I were to ignore the wealth of information that animals intentionally transmit, I'd hobble my own capacity to understand the world around me. But the only way for me to explain that world is with my human language, with all its limitations. I'd especially like to know the hog expression for "human."

When I realized that Indie was coming for me, I dropped the buckets and trotted toward him, meeting him halfway down the hill. He was chattering in his language, obviously distressed.

"Okay, Indie," I said to him in mine. "What's up?"

Continuing to chatter, he turned and started back to the others, still gathered at the fence. His message was clear: "Hurry up. We've got trouble."

He ran on and, following quickly, I rounded the bushes to find the problem: one of our goats, Gladys, had put her head through a square in the woven wire fence so she could eat the grass on the pigs' side. She couldn't pull her head out because her horns would catch the wire as she yanked back. She was stuck and had been for at least several hours; her hooves had dug deep into the dirt as she'd struggled and jerked.

The hogs looked at me. Then they looked back at Gladys, and then back at me again. Iris and Ruby shifted their considerable weight from side to side. White Foot pirouetted twice. Indie was still panting from his dash up and down the hill.

A conversation ensued. I acknowledge I've imagined the specific words of this conversation, but the hogs were intentionally using their voices and actions to communicate something about the scene in front of us. This translation is the best I have:

"Well! It's about time you got here. Aren't you going to do something?" Iris said in a tone that, if it had emerged from a human throat, would have been called indignation.

"We've been here for hours, protecting her," Ruby added. "The other goats wandered off. A coyote might have eaten her if it wasn't for us."

"I'm a bit dizzy," White Foot murmured, sitting down on her haunches.

Gladys yanked her head again with a weak, futile jerk.

"Can't we leave her now?" whined Indie, as he edged back up the hill toward his waiting dinner.

"Look," Iris commanded. "She's exhausted. Get her out of that fence!"

"Right," I said as I bent down to pull apart the wire and guide her horns back through the fence. In a minute or so, Gladys was free. Rushing back to her herd, she quickly disappeared through the pine thicket.

The four hogs seemed to utter a collective "harrumph" and turned away from the scene.

As I walked up the hill ahead of them, they rattled on, gossiping loudly to each other—maybe to me also, but maybe just to each other. I couldn't tell who was saying what, but I'm reasonably sure about what I heard:

"Those goats are idiots."

"And, you know, she didn't even say 'thank you.'"

"Maybe she thought we were going to eat her. You never know what those goats are thinking."

"Well, Indie did lick her face to calm her down. That didn't work too well."

"Yeah, she freaked out. But she's plum stupid. Why did she want that grass in the first place?"

"It's a good thing—what we did this afternoon. But I'm hungry. That human better have brought some decent food."

Iris got to the nice chicken leg first, and it disappeared quickly. All the troughs emptied fast. In their ferocious, slurp-slurp-slurping way, those four pigs devoured that meal as if it were the best food in the world.

I waited a few minutes to watch the post-dinner ritual. After the last morsel vanished, after checking and re-checking the troughs, after cleaning their snouts on their front legs, the three ladies ambled over to the nearest muddy spot and nestled themselves into the accommodating muck. They began gossiping again, discussing the afternoon's events or critiquing their dinner. Like I said, I don't talk hog so I don't know for sure.

My stomach growled, speaking a vernacular I understood well. I picked up the buckets, passed through the gate, and turned for a last glimpse of the foursome. Indie stood between the troughs and the other three, who were on the verge of dozing off. They stretched their huge bodies into a muddy bliss. Indie looked at me with a forlorn expression. Maybe he was saying, "You didn't bring quite enough food" or "I never get the good stuff" or simply, "I need my head scratched."

"Good night, Indie," I whispered and headed home to dinner.

The Duckness

Priscilla

Morning Ritual

We parked our chairs and table in the far corner of the screened porch, just inside from the lush, four-foot-wide viburnum. The scent of flowering trees and shrubs filled the morning air. I relished drinking my morning coffee on the porch in the late spring.

The viburnum and other bushes near the porch were loaded with birds and bugs, which made it a favorite spot for the other coffee-time regulars: our three, sort-of-pet ducks we called the duckness.

Henry and I sipped our coffee, let the caffeine hit our bloodstream, and chatted. When they heard our voices, the duckness paraded toward the porch in single file: Boss Duck leading, Mean Duck second, Nice Duck at the rear. Their prattle joined ours, and coffee time was in action—the morning ritual we all enjoyed.

The three ducks were male heritage Appleyard ducks, not the migratory green-headed mallards most people think of when they hear *duck*, but also green-headed. The farmer who'd bred them had let them get too old and large to butcher, but he didn't have the heart to kill them either. He wouldn't have cared if they had lived out their lives on his farm except that male ducks beat up on anything they can—including female ducks. Farmers try to keep just a couple of males (an heir and a spare), and female ducks make up the rest of the flock.

So, he offered his three-too-many male ducks to us, for free, when he came to pick up the goats he'd bought from us. Henry always commented on the way they moved as a unit, as if they were three expressions of a single mind. His tag, "the duckness," took.

I was halfway through my second cup of coffee and an explanation of my day's activities. As I finished describing my scheme for the day's events, I noticed a faraway look in Henry's eyes. He waved, as in "have a good day," and made his way to his office. His departure signaled it was time for me to jump into action.

As I cleared the breakfast table, I heard the duckness gorging on the wet bugs waking from the chill of night. Bon appetite, duckness!

I made my way through my morning chores in the house and then headed to the front porch to put on my boots. The ducks cackled all the way around the house as they came to meet me, waddling as fast as possible. They wobbled through the wet grass, grabbing each other, fighting to be first at the bottom of the front porch steps. I never discovered how they always knew I was headed there, but they were determined to beat me. Most of the time, I made sure they won.

Like every other morning that springtime, they arrived at the bottom of the steps in their green-headed, pompous way: single file, military strut.

"Good morning again, duckness," I said as I bent down to

stroke them. "Shall we start our day?" I said the same thing every day, and they always cackled together back to me, their heads bobbing up and down. We talked as we sauntered toward the front gate. I led, and they were right behind me, fighting each other for whatever position they wanted. They kept up their chaotic waddle and exuberant banter all the way to the gate. I imagined us as pals, laughing at each other's jokes while we saw one another off to work. I opened the gate and stepped into the pasture, closing the gate behind me. Once they heard the clang of the metal latch, the duckness did an about-face and marched back to the yard, all of which they considered their domain.

Their Dog, Dud

Sometimes in the morning I worked near the gate and watched the duckness track down our dog, Dud, and chase her around the yard. Even the cats seemed to laugh at the dog's humiliation. Dud was a Chesapeake Bay retriever and Newfoundland cross we had rescued a year earlier. She was broad-chested, but not heavy like Newfoundlands. Her hair was red but not wavy like a Chesapeake Bay retriever. Her teeth showed a pronounced overbite, the first thing you noticed about her face. Her name was an acronym of Dumb Ugly Dog. But we loved her, I guess. It was hard not to, because she was so clueless about almost everything. One of her adorable features!

When the duckness first met Dud, they were aggressive, running at her like three hardened warriors. They flapped their wings and stretched out their necks in attack posture, shrieking their war cry. Dud was so shocked she fled to the far side of the yard, yelping as she ran. The duckness followed in hot, screeching pursuit. Dud dove under the house to get away from the vicious ducks. The ducks danced outside her hide-away, bobbing their heads and squawking. When Dud didn't budge, they decided to move on to another adventure. After a few minutes of quiet, Dud crept out of her hiding place and crawled under a tree to sleep off her ordeal.

After that first meeting, the ducks knew Dud was their servant. She would follow them most of the day and nap close by.

I felt bad for the poor dog. She must have been embarrassed at being bossed around by a bunch of ducks. Especially the one called Nice Duck. Even now, when we talk of Dud, it is always poor Dud and the crazy duckness.

Garden Time

After lunch, I usually worked in our vegetable garden, harvesting what was ready, repairing what had fallen over during the night, and crossing my fingers for rain.

I counted on the duckness to be part of my gardening. They ate their weight in insects, especially Japanese beetles and stink bugs, and devoured weeds. Bless their hearts!

When I found a tomato plant that was covered in Japanese beetles, I yelled for the Duckness Bug Patrol. They came waddling as fast as their six webbed feet could take them. I shook the bush, the beetles fell to the ground, and the duckness had their feast.

The second most useful thing about them in the garden was their skill and delight at chasing squirrels. The squirrels would race to the trees where they could escape back into the woods for a while. Although I never did figure out what they had against squirrels, I was thrilled when they chased them out of my garden. The duckness was as good as a beagle!

Sophie

One day in late September, I gathered a group of goats in the field closest to the house and waited for a customer to pick them up. Sophie, our youngest guardian dog, had helped me bring the goats from the back field and was outside the pasture fence. There had been a lot of changes on the farm—as there were every fall—and she was on edge, not knowing why some of her goats were separated from the others. Like the other dogs, her mission was to protect everyone in the herd. More than once we had had to coax one of our dogs out of a customer's trailer because they had followed the goats inside.

She was walking back and forth along the fence, keeping one eye on the goats and one eye on the odd birdlike creatures that were bobbing around the front yard, not far from the pasture gate. The birds she knew well—eagles, ravens, and robins—were pests that threatened her herd. But these birds had green heads, waddled on weird floppy feet, and never stopped talking. They had feathers. Birds for sure, but strange. Possibly dangerous.

The customer was late, and I was just hanging with the goats. The warmish afternoon wore on quietly, and I noticed that Sophie was no longer pacing outside the fence. "Well," I thought, "she's gone to the woods for her walk-about." She'd be back at the sound of a truck pulling up the driveway.

But then I glanced toward the house. Sly Sophie hadn't gone to the woods: she was going to check out the odd bird creatures.

I watched as she sauntered toward the duckness, who were hunting bugs in the garden.

They noticed the strange dog but didn't think much about it, especially since their experience with dogs had been Dud, their servant. They kept plucking up bugs while also keeping an eye on Sophie, who suddenly came a little too close for comfort. This dog had a certain look in her eye they'd never seen with Dud. The duckness turned to waddle off.

But Sophie followed so she could smell them up close, get the lowdown on what they were. I thought they'd do the mean duck thing—heads down, wings out, and a fast run toward the attacker—and she'd back off. But Sophie had another idea. As the duckness picked up speed, she realized she was losing her chance to find out what these creatures were all about. Sophie opened her mouth and grabbed the last duck on the back of the leg. She held him firmly, not hard enough to kill him, but hard enough to let him know not to move.

Surprised at this sudden development, I watched in silence. If I yelled at Sophie, she might let go of the duck—or she might bite down harder.

Sophie turned from the other ducks, who'd decided to be mute, and proudly trotted to the fence. She was ready to take her strange new treasure back to the barn for show-and-tell.

In a gentle, firm way, she held the duck in her mouth. But when she arrived at the gate, she couldn't figure how to jump it with a duck dangling from her lips. For several minutes, she stood there—a motionless dog with a motionless duck suspended in midair.

I waited motionless, too, hoping my instinct to leave things alone was right.

I heard the other two ducks from where they were hiding behind some hay bales near the garden. Screeches. Squawks. High-pitched cackling.

And then, without any signal, Sophie carefully laid down the duck, backed off, and jumped the fence to rejoin her goats. She never looked back. I'm sure she had a great story to tell that night.

I was worried about Sophie's behavior but now was not the time to think about it. I had a freaked out, terrorized duck to look after.

The duck lay on his side. He never moved or made a sound, which is hard for a duck. I dashed to the gate to inspect his injuries. The two other ducks waddled to their friend in a noisy, frantic display of concern.

I wrapped the duck—it was Nice Duck—in a blanket and hurried to the bunkhouse. The other two ducks followed closely, watched as I doctored their mate, and commented on every application of ointments and bandages. I did all I could and then made a bed of hay that was big enough for all the duckness to sleep safely on the bunkhouse porch. They settled down. I put our wounded guy in a blanket. The other two ducks slept beside him in the hay.

The next morning, to my surprise, the blanket was still wrapped around the injured duck. I carried him outside and placed him in the middle of the blanket where he could watch his mates. Throughout the day, I put bugs and greens close enough for him to reach them. I was touched by the way Dud moved cautiously around him and lay at the edge of the blanket all day. After a few days, he started to limp around. We tried to keep him on the blanket, but sometimes he waddled away or just lost his balance and rolled off the edge.

During the day, Nice Duck's devoted companions took turns checking on him. For a while, the duckness slept on the bunkhouse porch and Dud lay outside the porch every night.

Although Nice Duck's wound healed and his confidence came back, he kept a pronounced limp. So, we renamed him Disabled Duck. When he got excited, he toppled over completely, but it didn't matter. He was still one of the gang. He still showed up in the morning under the viburnum, and he was still part of our family rituals.

Sophie never bothered the ducks again. That episode helped her learn that no matter how strange the ducks acted, they were part of the farm—and not to be eaten or used for show-and-tell at the barn.

Boss Duck, Mean Duck, and Disabled Duck remained a crazy gang for a few years. Three pompous green heads, single file, military strut!

Sex in the Pasture

Henry

"Look at the size of his testicles!" my twenty-something colleague hollered loudly, reaching out to clutch her boyfriend's arm. They'd just caught their first glimpse of Pedro, the largest and most impressive male goat we'd ever had on the farm.

"Those balls are huge!" she exclaimed. Her frank enthusiasm didn't bother me—farm life encourages plain speaking—but I wasn't sure about her boyfriend. He squirmed a bit and didn't say anything.

Priscilla and I were hosting a potluck picnic for about twenty of my coworkers and their families or partners. It was a beautiful cool mid-October Saturday and, before anyone arrived, I mentioned to Priscilla that at least some of our guests would want to take a tour of the place. None of them had been to our farm before and probably none of them had seen goats up close. Priscilla was the logical choice for leading the group since she knew far more about our pastures and our goats than I did.

"No way. I'm not walking around with your colleagues," she said. "No offense to them or anyone, but if they don't know about goats, they'll just ask all about obvious things that they could figure out for themselves if they just used some common sense. You know I can't stand that."

"But ..." I began to reply.

"I'm sorry. I don't have any patience for stupid questions." This I knew to be true. They just fueled her impulse toward nasty sarcasm.

All our guests had graduate research training in some area of health care. They were no dummies; some were trivia wizards. A few might even have known the location of the first wild goats: the Fertile Crescent, in the mountains of what are now Turkey and Iran. But my colleagues were what most farmers would consider book smart. None of them had any experience with agriculture, and certainly not with shepherding goats.

It'd be best if I led the tour.

About ten or so of my colleagues accepted my invitation to walk around the farm—past our pastures, goats, horses, and the pond. Our first stop was the buck pasture, the one-acre paddock where we kept the two or three bucks (that is, the male goats) who'd be our herd sires for that year. When breeding season arrived, usually in early to mid-November, we'd split the doe herd into two or three groups, and let one buck mingle with a specific group, depending on our breeding goals.

We were getting close to the day when we'd introduce the bucks into the pastures with the does. Some of the does were already in estrus, and the bucks could sense it, even though the doe and buck pastures were separated by several hundred yards. As a result, all the bucks were walking the fence line on high alert. Acting like the king of the pasture (which he was), Pedro strutted around with his pendulous testicles swinging back and forth. He'd occasionally rear up and butt the other bucks just to make it clear that he was in charge. Although they were strong themselves, they stayed out of his way.

My coworker's enthusiastic observation was scientifically accurate, as I told the others in the group. Male goats generally have large and obvious testicles because evolution has favored comparatively large testicles in mammalian species with multi-male breeding. In species of goats and sheep, different males will mate with a single female if they are left in the same pasture during the same estrus period. Over the long haul, males with larger testes win the competition because they have higher sperm production and can copulate more frequently than males with smaller ones. At least that's the theory.

What's obvious in practice is that full-grown, intact bucks, especially impressive ones like Pedro, can mount many willing females—fifteen to twenty in a day's time—without any apparent fatigue. And, for the record, a buck will not mount an unwilling female. He may sniff around, but if the female isn't receptive, the buck backs off. There's no rape in the goat world. I hadn't planned to talk about any of these details during the tour, but a few of my colleagues kept asking questions. I was glad Priscilla wasn't there.

Anyway, the group finally wandered on to less exciting places. The varied colors of the does in our herd—black and tan, white and cream, pure black—made for a pretty scene against the mountain, itself spotted in a bit of early autumn color. The horses obliged by coming over to the fence for carrots. And we stood at the pond's edge for a while, watching a cloud's reflection drift silently across the water's surface. For many of my colleagues, being in the country was a liberating relief from the city's restless thrum.

We turned homeward, looping back to the road that ran past the buck pasture. I was walking in the middle of the group, talking with someone—most likely about some problem at work—when I again heard my colleague's loud voice. She was some distance in front, but her words were clear.

"Oh, my God," she hollered. "Look at the size of his penis! He's peeing on himself! What is going on?" I caught a glimpse of her boyfriend rolling his eyes. He tried pulling her away, to no avail. She was glued to the fence, watching Pedro do what bucks do.

I probably should have anticipated this reaction during my earlier explanation of bucks' testicles, but I didn't think I needed to cover every aspect of a goat's reproductive behavior. Now she was witnessing the common practice of male goats' self-urination.

The big boys spray themselves with their own urine to attract does with its musky odor. Their aim is quite accurate. They can douse their front legs and chin with precision because they have a penis that's about 11 inches long, plus or minus, and about an inch across. When it's fully extended from its internal sheath, it reaches far forward under the goat's belly, giving it a wide range of motion—up and down, side to side—that bucks seem to control easily. Pedro's penis probably exceeded the average length because everything about him was extra large. I hadn't measured it and wasn't going to. Let's just say that my colleague wasn't the first person to be impressed.

A musky stench drifted from the buck's pasture. The group moved along, some walking faster than others, and we joined the rest of the crowd at the house.

At some point, the folks who took that tour must have decided never to ask me about our goats again. For the many years I worked with them, the subject never came up. Maybe they'd learned all they wanted to know. Or maybe they were uncomfortable with how the primitive and the wild complicated their idea of a farm. It's true that our goats are domesticated livestock. They live in pastures and allow humans to manage them. But when it comes to sex, they are still wild, still driven by the same primitive instincts as their ancestors that roamed the mountains of the Fertile Crescent 11,000 years ago.

Hat of Shame

Priscilla

As I got the coffee going in the kitchen, I glanced out the window.

"Oh, shoot!"

One of my young Spanish does was dancing in place with her head stuck in the fence. And, wouldn't you know, it was the line of fence that backed up to the house of one of our neighbors, whom I like to call Dastardly Darlene. I was certain she'd been watching the frantic goat from her mudroom windows.

Sure enough, at eight o'clock on the dot the phone rang. I picked up the phone, and before I could get it to my ear, Darlene screeched in that high-pitched, exaggerated Southern drawl she loved to use on me. "I'm callin' the sheriff this time, Pris. I've had it with your neglect of these animals. One of your poor goats had her head stuck in the fence probably the whole night. She's been yelling since early morning when I let my Bear out. I wouldn't be surprised if she can't breathe—or maybe she's dead!"

Darlene always amazed me, the way she could raise the last word of her tirade at least an octave to punctuate her southern-ness and to emphasize her disapproval of my farming practices.

"I'll get to it, Darlene," I said in my most diplomatic way. I tried to hang up the phone, still hearing her voice.

"I'm not kidding, Pris!" she screeched one more time.

"I hear you. Darlene." I hung up the phone gently.

I took my time finishing my coffee, then pulled on my boots, grabbed my leather gloves, sprayed my hat with bug repellent, and shuffled out the door.

In the truck I knew I'd better check for duct tape and short, thick sticks—a goat farmer's go-to tools for keeping goat heads out of wire fences. I drove to the field and pulled up to the gate, and there was Dove in all her glory. She'd caught her head in the fence and was having a fit.

Arnost, the guardian dog, seemed in a nasty mood. He hovered beside the distressed goat, barking and growling. Dove swung her head up and down and back and forth, bawling as loud as she could.

Something on the other side of the fence was antagonizing Dove, but I was too far away to tell who or what it was. I grabbed my stick, tape, and wire cutters and started across the pasture. As I grew closer, I saw the problem: Darlene's annoying little dog, Bear, was jumping with glee, yelping, and licking Dove's face. He was having the time of his life torturing my goat.

Dove pulled back and lunged sideways away from Bear, straining this way and that, using extraordinary, never-before-seen movements. But she could not free herself from this face-licking maniac.

Bear continued to jump and lick, jump and lick. A strange, high-pitched whine came out of him as his dance became more frenzied. He was in ecstasy, enjoying his freedom to be hateful and mean. Dove's goal was surviving the torture.

Arnost walked slowly over to the fence and looked at the yipping, crazed little dog.

This was not going to be pleasant. A captured goat, my guardian dog in a bad mood, and one demented, creepy little dog driving my goat into a frothy mess. I was sure Darlene would show up soon to top off the circus.

Dove screamed with renewed agony and twisted her body trying to get her head free. A quick glance at Arnost, and I could tell he was hatching a plan to take out this demented little dog on the other side of the fence.

I had to work quickly. Getting Arnost to back off was my first job, but that would not be possible until that whirling, out-of-control Bear was out of the picture.

As if on cue, Darlene exploded through her mudroom door. I was not surprised, because she'd been watching everything, calculating when, exactly, to make her grand entrance.

"Thank heavens I have a sweet little dog and not one of those vicious breeds like yours." She swung her hips as she strode toward us. She even got in some painful hops walking on bare feet. "Mama's comin', my darlin'," she yelled to her pooch.

She continued to chatter nonsense as she hopped across the yard in a knee-length pink nightgown highlighted with red roses, squealing when she stepped on acorns and sticks. With full makeup and her hair sprayed into the same hairdo she'd worn since high school, she looked like she was ready for action. It was all I could do to keep from snickering as I watched her make her way to the scene of the calamity.

She finally reached the fence and pulled her dog off my goat's head. Bear wiggled in Darlene's arms, but she quieted him with a treat. Then she slowly rolled her eyes towards me.

"I can't believe it took you this long to get out here and help this poor animal," she drawled. "I've spent most of my morning worrying about this whole mess."

"I never meant to inconvenience you," I said as civilly as I could.

I suffered through Darlene's constant blabbering as I worked to pull Dove's head out of the fence. Snipping two wires, I was able to get her horns free.

"Hmmm," I said in response to Darlene's continued chatter as I swung my leg over Dove's neck to make sure she didn't run off.

Bear wiggled again and barked, outraged that his victim was being released from the trap.

"Hush, darlin'," Darlene cooed to Bear as she cuddled the idiotic dog in her arms. "You're safe now. Mama's got you."

Arnost was watching. His low growl grew louder. He seemed concerned about his goat but confused about the whole situation. Did he want to kill that crazy dog? Or did he need to save his goat? Whining took over the growling. But he kept his cool.

With Dove out of the fence, I held her head and straddled her neck. When I felt I had a good grip, I picked up the duct tape and stick. Now, I just needed to create The Hat of Shame—my magic weapon to keep goat heads out of wire fencing.

I held the stick across her horns and wrapped the duct tape around them and the stick, crisscrossing back and forth, securing the stick to the horns. The hard part was doing it with a goat that was more than a little disturbed under me. Rip, twist, press down. Rip, twist, press down. I wrapped until I could not move my creation.

My masterpiece, her hat of shame, was done! The stick was

too wide to go through the wire fence, and it would take enormous effort to get that contraption off her head.

Arnost stood there looking at me, then looking at Dove, then looking back at me. He was confused. Dove was wearing a strange, goofy-looking hat. How did his goat come out of this trauma with a new head?

Dove stared back at him wondering what he was looking at. Why did Arnost, her protector, back away from her with a worried look? They had just gotten out of a tough jam. They should walk back to the herd proud to have escaped.

The new weight caused Dove to sway like a drunken sailor. The wobbly goat with the big thing on her head convinced Arnost that this animal was not Dove, but rather some kind of replacement pretending to be Dove. Being the sensible dog he was, Arnost ran away.

Dove ran after him, bleating what must've been goat for, "What's your issue? Wait for me!"

By this time, Dastardly Darlene decided to leave the show and carry her dog back to the house. She complained in her exaggerated syrupy accent about my lousy farming habits as she hopped across the yard in her bare feet. When she reached her house, she turned and faced me, her flowered gown blowing in the wind. "You better make sure that fence is fixed on my side. I still might call the sheriff, Pris!"

I made sure the fence was mended and made my way back to the truck. I snickered as I watched Dove chase Arnost up the hill and into the next pasture, continuing to bleat for him to wait, while Arnost kept up a fast, strong trot away from the strange creature following him. The strange creature that once was his goat was still pursuing him.

What a way to start a day!

The cool morning air and the symphony of birdsong erased the obnoxious sound of Darlene's voice and her irritating dog's yelping. My vision of Arnost and Dove running up the hill gave me a good chuckle as I drove the truck around, doing various chores.

From time to time, one of the other does in the herd ran from Dove, and, of course, she ran after her. But by the end of the day, her best friend finally stood close to her and allowed Dove to graze nearby.

For months, the hat of shame worked its magic on Dove. However, with the advent of breeding season, the hat, although successful, had to come off. I just hoped Dove had forgotten her old habits of sticking her head through the fence. But, just in case, I collected a cache of sticks and stocked up on duct tape.

From Power Take-Off to Artificial Intelligence

Henry

To farmers, the power take-off (PTO) on tractors is a deceptively simple device that transfers power from the tractor's motor to a range of farming instruments. It's been my dependable friend for more than twenty years. Its shaft turns and makes my other machines spin, cut, and dig. Through all my years of farming, I couldn't have done what needed to be done without the PTO.

Like all its mechanical siblings and electrical cousins, the PTO is not part of the natural world, but it is part of the farm—as present in our lives as the sun, rain, and livestock. And, like that livestock, it sometimes has a mind of its own.

One day, I rented a large, worn-out contraption, often referred to as a "buggy," which farmers use for spreading powdered lime on their fields (which, in turn, lowers the soil's acidity and prevents poor-quality grasses from growing). To some people, buggy connotes a covered wagon with wooden wheels pulled by horses; to others, it's a grocery shopping cart. The vehicle I rented was a tall metal box on rubber wheels. The inside walls sloped downward toward a conveyer belt. When the wheels rolled, the belt deposited powder onto a rotating plate that, in turn, flung the powder onto the ground.

My buggy worked like a truck depositing salt on snowy roads—except it had to be pulled by a tractor. And rather than throwing salt on slippery roads, it distributed powdered lime on an acre of ground peppered with many old tree stumps.

Early that morning, I arrived back from the seed store with the buggy, unhooked it from my truck, and attached it to my tractor. Like most pull-behinds, the buggy had a metal arm with a hole that sat over the tractor's hitch. I slipped a large pin through the hole, securing the buggy to the tractor.

The buggy also had a shaft that had to be fitted onto the tractor's PTO to provide the power for the conveyer belt and rotator plate. And that's when I discovered problem number one: the buggy's shaft, extended fully, barely reached the PTO's stub. As I pulled it into place, there was no length to spare. I checked several times to make sure both ends of the shaft—with their rusty universal joints—were locked on. One end was on the PTO; the other on the buggy's drivetrain. It was tight.

I needed Priscilla to drive the tractor because the young man at the seed store said the buggy's distributor flap sometimes closed unexpectedly. That meant I had to walk alongside the buggy, making sure the flap stayed open.

The meadow awaited us with more than fifty stumps poking up in the morning sun. Pris started the tractor and put the PTO in gear, keeping the throttle low. The shaft turned, the buggy shook, and the spreader plate rotated—but very slowly. Not enough to distribute the lime any distance at all.

"Okay," I hollered, "rev it up." She pulled on the throttle, gently at first, then harder. The PTO turned faster and faster, the shaft between the PTO and the buggy turned faster and faster, and then problem number two appeared: the rotator wasn't turning at all. I heard an ominous slapping sound, like baseball cards flapping against the spokes of a bicycle wheel peddled by a ten-year-old in trouble.

"Kill it!" I shouted to Pris.

She turned off the motor, and the shaft slowed to a stop. I inspected the connection between the shaft and the buggy's drivetrain. The splines were worn so thin that the universal joint was barely catching them. Metal was slipping over metal. *Why did the seed store think this buggy was rentable?* The only way to get the buggy's drivetrain to move was to ensure the shaft turned at just the right speed. Too slow and there wasn't enough power. Too fast, and the universal joint would slip—flap, flap, flap.

We started again. Pris put the PTO in gear and, with some coaching, found the sweet spot: enough throttle to make the spreader throw out the lime, not too much to make the shaft slip over the splines. We entered the meadow, lime dust swirling around me as I kept an eye on the rotator. Sure enough, problem number three showed up quickly: the flap that let the lime drop onto the rotator had closed.

"Hold up," I yelled. The tractor stopped. I pulled the lever to open the flap again and tightened its butterfly nut with a few hammer blows. Hammering on the nut meant the flap would stay open. *How often am I going to have to whack this thing?*

"What's wrong?" Pris hollered over her shoulder. "You know I've got to dodge these goddam stumps, don't you?" She waved at a couple of stumps dead ahead.

"Nothing's wrong." I whacked the nut one more time. "Okay. Get going."

Our repeated movements charted the choreography of a classical ballet: tractor weaving deftly between stumps, buggy lumbering elegantly under a cloud of lime dust, man whacking gracefully on buggy's wayward flap. The dance even had its own musical score: purr, whirr, thwack. Purr, whirr, thwack. Purr, whirr, thwack.

And then, problem number four arrived. At the meadow's end, the tractor made a U-turn and headed down another lane. As I watched, the buggy caught one of the stumps, ran up on it, and leaned away from the tractor, tilting like an ocean liner listing in a giant wave.

Pris jammed on the tractor's brake. Too late. The shaft pulled off the buggy.

But the shaft was still attached to the PTO. And the PTO was still turning fast, which meant the U-joint on the PTO's end of the shaft was still turning fast, which meant the shaft itself was waving and flying and bucking through the air like a colt who's stepped into a mess of yellow jackets. The classical ballet had turned into a wild modern dance, complete with a Philip Glass–like score of bangs, thumps, and knocks.

I'd never seen a run-away shaft before. It was twirling, pirouetting, wiggling. A herky-jerky harpoon wielded by a madman.

I started to laugh.

The tractor, still connected to the buggy with its metal arm, shook from the shaft's antics, rattling Priscilla. She looked back. The buggy almost toppled over.

My chortles become loud guffaws.

Pris had a white-knuckled grip on the steering wheel, and

her straw hat shook in time with the flapping shaft. Her face was pure terror.

I was lost in rolling waves of laughter, a tsunami of mirth, trying to find some way to say something. In vain.

"Waddaya want me to do?" she hollered. "Back up or what?" The shaft vaulted violently upward, dove down, whacked the stranded buggy, and leapt up again. The buggy tilted a bit more as the load of lime shifted, sending up a puff of dust. The tractor shimmied.

I grabbed feebly at my ribs, hurting from all the laughter. I couldn't answer her, which rattled her even more. The tractor's rumble drowned out her swearing.

Finally, she took the PTO out of gear. The dance ended abruptly. The shaft fell with a concluding thud. I was down on one knee close by, out of breath from hooting.

"You could have gotten really hurt," Priscilla screamed, glowering down from the tractor's seat. "That was a close call!"

"I know," I gasped. And then, as I pictured the shaft's wild dance, my rib-bending laugh began again. No way to stop. My eyes teared up. I glanced at Priscilla's worried face and convulsed again.

Finally, with considerable effort, I suppressed the sniggers. "Just back up slowly," I hollered, wiping away a few tears.

As the tractor pushed the buggy backwards, it came off the stump and righted itself. I re-attached the shaft, barely able to see through my tears. We finished the job, but I couldn't look at Priscilla without guffawing.

Later, after I returned the buggy to the young man at the seed store and suggested he take it to the dump, I thought about what might have happened if I'd been standing closer to the buggy's front end when the shaft had pulled off. It could've clubbed me hard. A bad bruise? A broken leg? Worse?

The PTO itself was blameless, but these things—machines acting unpredictably because of unforeseen problems—happen often on farms. A chainsaw jumps back and slices a boot; an old tractor slips into gear and runs away because a shirttail catches a throttle; a backhoe bucket somehow wraps around a barn's corner post and won't let go.

What about all the electric tools in our lives: smart phones, smart TVs, laptops? We depend on devices and technologies that we barely understand. They help in uncountable ways, but disaster often hovers: a stolen identity, a mysterious shutdown, a sudden discovery that someone unknown can listen to your kitchen conversations through a small speaker. These tools are like friends with a questionable past—we trust them at our peril.

And now, how are we to understand computers that run systems of artificial intelligence? Many AI experts believe that continued, unprincipled development of such systems could modify life on this planet profoundly. A dangerously unpredictable, out-of-control, PTO-driven shaft that we can shut down with a simple action is one thing; a dangerously unpredictable AI system that can't be turned off (and won't turn itself off) is another. Neither an AI expert nor a buggy-man, I'm nonetheless sure that part of our future with AI will be no laughing matter.

Hard Times

Pedro

Priscilla

It was still dark when we crossed the Mississippi River near Cairo Junction, Kentucky. My son was driving our truck, towing an empty stock trailer. Two hundred miles across the bridge in Joplin, Missouri, was Pedro, the Spanish buck we were traveling to pick up. A thunderstorm had dogged us for a while, and now it intensified. As we rolled onto the bridge, we saw tornadoes on both sides, undulating grotesquely in strobe-like lightning.

The rain had been falling in sheets. Now those sheets blew sideways, sounding like pebbles pounding the truck. Sandwiched between two eighteen-wheelers hurling themselves over the bridge, we surfed on the waves from their front bumpers. All I could do was stare forward in utter terror. Finally, I found some voice.

"Are we alright?" I asked my eighteen-year-old son, fresh home from his first year of college. He had agreed to help me with my mission.

"Sure, Mom," he said. "No problem."

I reverted to mute and stared out the window.

After we crossed the bridge, we made for the first exit. My son pulled the truck and trailer into the closest truck stop. The rain had eased up, and the winds had calmed. I breathed deeply, a sigh of relief. We were lucky to be across the river and in one piece.

"That wasn't so bad," my son said, with a flushed, amazed look on his face. "You okay, Mom?" he asked.

I'm sure I looked exactly as pale and shaky as I felt.

I glanced back toward the Mississippi where the storm was still raging, traveling up the river. My son and I agreed that the tornadoes had moved away. At least that's what we decided to tell each other. I hoped it was true.

We got some water and stale sandwiches from the store and ate them quietly in the truck. Nick finished, and, exhausted from the white-knuckle driving, said he needed a nap. I drove the rest of the way.

The next morning, we arrived at the farm to pick up our precious cargo of five special goats. We told of our harrowing trip, had a good meal and some sleep, and were back on the road, heading home to West Virginia.

Pedro and the four does with him were definitely worth the trip. He was huge with a rare tricolored coat—dark red, dirty brown, and black with white markings breaking it all up. He stood with an assertive stare and crowned with a massive pair of horns. His head was handsome, and I had a hard time keeping my eyes off him. He was just flat out impressive.

We only had to drive through a couple tornadoes to get him. I smiled, thinking of the discussion with Henry right before the trip: "Free goats?" he'd exclaimed, trying to control himself. "You said you'd take on these goats?" He stared at me with the look that made me know I better have thought this out. "You are going where to get these free goats without even seeing them?"

He knew that when it came to building a goat herd, there were some things I just did not do because they were far too risky. He also knew that I was about to do one of those things.

"This will give me some really important DNA in my herd," I'd said, looking at him with eyes full of as much conviction as I could muster. He was well aware that I always tried to have a well-thought-out breeding plan for my herds of Spanish and Savanna goats. Bloodlines of heritage livestock don't have the forgiving numbers of commercial animals and so I was always terrified of making too many mistakes.

"Pedro's bloodline is necessary, and he will cover all my important true Spanish does," I said.

There was silence between us for a minute or two. "And you're going where to get these free animals? Missouri?" He asked, as he paced around the room. "We are in West Virginia and Pedro is in Missouri, which means going into tornado alley in peak season and crossing the Mississippi River just before dawn, at best," Henry exclaimed, running his hand though his hair as he tried to take in my plan. The hand in hair was a signal of great frustration and worry. "And you said, no problem, to this person?" With that question, he walked out of the room.

Pedro settled down on the farm quickly and was soon Big Boss. He was happy with his herd of does and everyone seemed healthy. I could hardly contain my excitement for the new kid crop that he would sire. My grand plan seemed to be working! Henry even seemed to realize how special Pedro was.

One day in the late fall, the alarm rang and I hurried to silence it. Henry had been off the farm for over a week. He'd returned last night, tired and glad to be home; I thought I'd let him sleep. I lingered in bed for a few moments, enjoying the brisk air streaming through half-opened windows and my husband asleep beside me. As slowly as I could, I raised off the bed and searched for my slippers.

Behind me, a low soft voice. "Don't go yet," the voice said. Turning around, I saw my husband's sleepy smiling face. I slipped back under the sheets.

The whole morning got off to a slow start and I decided to keep in that rhythm. I moved through the barn, wrestling with new bags of feed and filling the buckets for Pedro's herd. The walk to the pasture was long, but very pretty. I could see the gate in front of me.

The first strange thing I noticed was Arnost, the guardian dog, crouched at the gate with his tail between his legs, whining. Looking past him into the field, I saw the herd. They were in a tight group, near the other corner. I set down my buckets and opened the gate. None of the does turned to face me as I came into the pasture. Instead, they kept looking forward as if they were just staring into a far away space. That was very strange.

A cold rush ran through my body. Walking cautiously into the field, I looked in the direction the goats were looking. And there it was: the mangled body of Pedro. It was suspended in the fence's wire, his head twisted grotesquely, his eyes staring straight back at me. He had been electrocuted.

His magnificent horns were tangled in the remains of the ripped fence that now netted his complete body as if he were caught in the web of a giant spider. His back legs lay against the ground. He must have fought death to the point where the fence wire was torn apart but he could not free himself from the relentless current. He had become a tragic display of the struggle to live.

In a daze, I turned off the electricity and returned in complete horror to gaze into Pedro's deep hazel eyes with their black horizontal retinas. In life they were always darting and dancing. Now they were still and cloudy, covered in a dull film. His front legs hung, folded in unnatural shapes and wrapped with wire. They seemed delicate, dangling in space. When I touched them, they were stiff and cold.

I just couldn't believe what I was seeing. I had no voice. I kept thinking I was wrong, that it isn't him but one of the other big does with massive horns. Or that I'd be able to get him out of the fence. Or that he'd wake up, sore but able to limp back to his herd.

I tried to pull his legs out of the fence, but I had no tool to free the lifeless body.

I made myself turn away. At that moment, I knew this was real. Pedro was dead.

With a deep sigh and wiping away my tears, I switched, as always, to trying to think of the work that needed to be done now. First, get Pedro out of the fence and then patch it to make sure his herd was safe before dark.

I walked back to the house, looking for Henry, whom I found in the garage. I heard myself asking him for help, but I felt like I wasn't the person who was talking. I must have looked sick or faint because suddenly he was holding me. Slowly, with his help I steadied myself. We sank down on an old sofa, and I told him of Pedro's death. We worked out the jobs in front of us. But then, I passed the point where I could hold in my emotions. I dropped my face into his chest and cried and cried and cried.

He held me as I wept—for Pedro, for the does, for myself, for the farm, for my dream of this year's kid crop. Henry held me until the tears ran dry and I had had my fill of misery. It was now time to get on with the things that had to be done before dark. It took a few hours to get poor Pedro out of the fence. Then, we had to move the goats to a new pasture. For once, they didn't give us trouble. In fact, they went politely through the gates. Arnost realized he still had to take care of his does and came with us quietly. He didn't even run to the woods for his usual romp.

Henry and I looked for a good level spot in the forest behind the far pasture, a place that could accommodate a farm cemetery. Pedro wasn't going to be the last goat we would bury. After selecting the spot, we loaded Pedro into the tractor's bucket. I walked behind as the tractor bumped over the uneven ground, Pedro's huge horns sticking out from the bucket, swaying with each bounce. I kept imagining he'd jump out of the bucket and head straight for his does.

I could not watch Henry bury him, so I sat on the nearest fallen tree and waited, digging into the ground with the toe of my boot. We rode the tractor home together in silence.

I still needed to figure out what the hell had happened. Also, what was Plan B? Who would take Pedro's place for the rest of the breeding season?

Henry repaired the mangled part of the fence as I looked for clues to the mystery. I noticed that a young Savanna buck had somehow gotten out of the pasture he was supposed to be in and was grazing in the pasture next to Pedro's. He was limping and had blood on one shoulder.

Then I knew what had happened! The boys were fighting over girls! Oldest story in the world! This Savanna buck was looking for relief and I guess cocky enough to think he could fight Pedro for his does. Pedro's horns and legs had become tangled in the fence as he tried to attack the young Savanna buck on the other side.

I drove the buck into another, more secure, paddock and away from Pedro's herd. "I've had enough of you for a lifetime," I yelled at him, as I locked the gate. I wished I'd buried him and not Pedro.

Now on to Plan B. In farming you always must have a Plan B. We had to put another Spanish buck in with Pedro's group of Spanish does and we had to move fast. Breeding season was coming to an end. The first young buck lasted around thirty minutes; he climbed over the half-door of the barn to escape the biggest does, who were battering him like a punching bag.

The second young Spanish buck was not much bigger than the first, but he was stocky and agile. He came in strong and aggressive to Pedro's group. The does didn't know how to take this new approach, and he seemed to win the day. At least he wasn't trying to escape. The next morning, he seemed to be comfortable with the herd. I let nature take its course. That season the young buck earned his name: Bulldozer, a name he was given by the woman who bred him.

In early spring, we watched the new kids be born. We were pleased to see that many of Pedro's descendants were among them. There were some real beauties. Even though his reign as herd sire had been short, he had covered most of his does and we had a nice crop of his offspring.

And those offspring tended to be big. All of them—even the doelings—had their dad's impressive horns. Many of the bucklings had their father's hard stare and his strong beautiful hazel and black eyes. We also had quite a few kids with a tricolor coat and long legs.

Bulldozer's offspring carried his black and brown coat, and their horns were closer to their heads and not so massive. He fathered a lot of does, which was a big plus for my breeding program.

Pedro's children went on to influence my herd and more. One of his bucklings looked just like his dad: same tricolored shaggy hair, same dignified stare, same set of huge horns. He became a great buck. After using him for a couple years, I sold him to a university studying Spanish heritage goats. I'm proud that Pedro's genes lived on in the development of the Spanish goat breed.

For many years, when I looked at the herd in my pasture, I was able to pick out Pedro's offspring. They had his majestic head and stood far taller than most others in the herd. In the evenings as I walked back to the house, I imagined him towering over his does. He will always be my favorite Spanish buck. I will always be honored to have had his bloodline in my herd. I loved him. I still miss him.

Death on the Farm

Henry

Close-Up

During our years of farming, Priscilla and I became more familiar with death—its particularity, inevitability, and randomness. Death took many forms. Experienced does sometimes delivered babies stillborn. Small bucks went to the butcher. Two of our guardian dogs died with little warning from causes still unknown. Some animals died from infections, others from accidents. And sometimes—more often than I wished—I was death's agent.

In television shows about nature, viewers watch predators stalk and capture their prey. We typically witness what the narrator assures us is a normal process of life in the wild, but the sudden violence of death happens at a distance to animals we know only as a species. Even most movies and television shows about farming sanitize death or treat it as a minor blip in a farm's ongoing flow.

On our farm, death was close-up, messy, personal, and constant. Because Priscilla spent so much time with the herd, she knew each goat as an individual and as part of a family lineage. She traced the ways that grandmothers, mothers, kids, cousins, and aunts stayed together and cared for each other; the family bonds were obvious to her, which made each goat even more distinctive.

We couldn't afford vet visits every time an animal was sick, but Priscilla was good at diagnosing our livestock's health problems. She'd grown up around animals all her life, read books about large-animal care, talked to many other farmers, and listened closely to vets when they did come to the farm. She'd become knowledgeable about common illnesses in goats and treatment options. As a result, the one-year survival rate of our newborn kids was above 90 percent and we rarely lost a goat to a runaway infection. When one did die, it was a personal loss, bringing at least a twinge of guilt. Maybe we should have done more or caught the problem earlier or tried different meds. Inevitable as it may have been, the animal's death felt like a failure.

Whether our animals died through design, age, or accident, their distinctiveness could not be replaced. We could no longer witness their particular antics. Their contributions to the farm slipped into history. Every death also stole a story from our future; that animal's character would never be revealed in its actions. We can't know what those untold stories might have told us about life.

Our animals were precious to us not only because they were a source of income but because they deserved to live a decent life and die a decent death. As their custodians, we were responsible for giving them that respect. It's a straightforward philosophy with complications that we didn't anticipate.

The first year we had our farm, I had to shoot a young doe, whom we had named Brenda. She was thin, dying from a load of parasites resistant to our dewormers. It made sense to spare her the agony of a slow death. If our herd were wandering in the wilderness, we reasoned, it would have left this goat behind, and predators would have finished her off.

But that rationale was weak. Our herd wasn't in the wilderness, and we had dogs to protect our livestock from coyotes. The fact was, we didn't want her to suffer pointlessly, and we didn't have the resources to pay our vet to put her down by injection. Of the choices available to us, shooting her was the least expensive and most humane, if you can call it that.

I used my high-powered hunting rifle to kill the young doe. Although she died instantly, the death was messy and ugly because the bullet was big. I suppose it doesn't really matter how she looked once she was dead, but I felt bad about it anyway. I used my tractor to dig a grave deep enough so the coyotes couldn't get to her. It was a mossy spot behind the back pasture with a nice view of the valley.

A few years later, Polly, one of our oldest does, became sick soon after she gave birth to twins. By that time, I'd put down several other does and had developed a routine with my .22 that was as quick and dignified as I could make it. The smaller bullet left only a little mess and did the job just as finally. Polly didn't have her usual energy and her eyes were whiter than they should have been. We treated her for intestinal parasites. She got better, but not much. She never regained her weight, and she wasn't able to mother her kids very well. We guessed cancer was to blame, but without an extensive work up, we wouldn't know for sure. After she weaned her kids and continued to struggle, we decided to put her down.

I buried Polly near several of the other goats I'd killed. Their graves were in a flat spot over the hill where no one would notice.

No view to speak of. My need to make it matter of fact measured my difficulty with the act of killing.

Tiny Tim

When Priscilla told me that Tiny Tim needed to be put down, I replied instinctively, "No. He couldn't be that bad."

But she was right. When I went to check him out, he was in bad shape—dehydrated, thin, white eyes. He'd stand up for a few minutes and then slowly fold himself down on the hay.

I did not want to put Tiny down myself, so scheduling a farm call for the vet offered an alternative. But bringing in a vet was more than the farm could afford. Because his parasite load was large and persistent, he was a threat to our other goats. And he wasn't going to improve. If I was going to be the one to kill him, I'd have to do it soon.

Tiny wasn't the usual goat and the whole matter ushered in some knotty questions. Wasn't I letting myself off the moral or emotional hook by paying someone else to do the discomforting job? Was he in fact a commodity whose worth was so little that the cost of a vet couldn't be justified? Was an injection truly better than immediate death from a bullet? I had no clear answers to any of these questions.

"We've told each other that we wouldn't let our livestock suffer needlessly," Priscilla reminded me as I loaded the .22 with a few bullets.

"Yes, I know. It's time," I said.

"I'm sorry you have to do it. I just can't watch. Not with Tiny."

"Right. It's okay."

When I got to the barn, Tiny was standing in his stall, as if he wanted to join some of the does grazing nearby. I put a collar on him and led him to the side of the barn where there were no goats. He lowered his head to nibble some grass. I stood to the side and watched for a few minutes.

Tiny raised his head and looked around for his herd mates. I raised the gun barrel, placed it next to his ear, and made the shot. He rose on his hind legs, squealed loudly, and fell to the grass with a thud. A few leg kicks. Then, he was gone. It was done.

I had put rifles to the heads of other animals—old goats dying of cancer, a barn cat with multiple injuries from a dog, deer wounded in the hunting season—and each time, before pulling the trigger, I forced myself into a dispassionate, clinical state of mind. I would become a different me for a few moments, a not-me me. The mental trick never worked very well, but it got me through. Anyway, the worst part was not pulling the trigger. It was witnessing the inevitable, involuntary jerks and shudders of a small body as its nervous system shut down, as its life left this world. A tick of silence always seemed to follow the instant of death, as if the beat of the universe had paused momentarily, as if I were standing still in the space between before and after.

But life is indifferent to death. It goes on, relentlessly. I turned away from Tiny's body, grimly determined to remain matter of fact as a barricade to a spreading sadness. "I'm sorry," I said out loud, but no one was there to hear me.

The Day the Duckness Died

The trio of our Chaplinesque Appleyard ducks arrived in August. We soon named them "The Duckness" because they always went everywhere together, usually in single file, as if they were three

Chaplins parading in some park, swaying slightly side-to-side, listening to their own internal music, and seemingly entitled to make trouble.

In November, the group was hobbled somewhat when one of our dogs clamped down on one of the ducks, leaving its left leg broken and a bit bent. The duck's disability was a minor nuisance until a bitter cold day in January when the ducks spent the night near the hot tub, which still held warm water from our tubbing earlier in the evening. They settled down in the space between the house and the tub, which was damp but mostly clear of ice and warm from the tub's radiated heat.

The next morning, I put on my heaviest coat—the outside thermometer pointed to the seventeen degree mark—and went to check on the duckness. Two of the ducks had wandered off a bit but were calling back to the third, the disabled duck, who was still sitting in the spot he'd occupied the previous evening. I walked over and bent down. He flapped his wings in obvious distress. I tried to lift him up. He flapped his wings faster. The other ducks came waddling over, quacking loudly, which I interpreted to mean "Don't touch him! Let him go!"

As I tried to move his body, I realized he was stuck—frozen into the dirt. At some point in the night, the ground's moisture or maybe water leaking from the tub had solidified so quickly that he was caught. The others had had enough awareness and strength to pull themselves off the ground or change positions during the night. This duck didn't and if he wasn't freed, he'd be stuck for a while. The forecast called for at least two days of below freezing temperatures.

For reasons I can't remember, I began calling him "Chester."

Maybe the name sounded like the right combination of reassurance and urgency.

"Okay, Chester," I said to him. "I'm going back to get some warm water so we can melt the ice. Tell your buddies to calm down."

When I returned with a pot of warm water, I was startled to see one of the two able ducks standing on top of Chester, his two large, webbed feet massaging Chester's back. I watched for a moment, trying to understand what was going on. The duck finally hopped off and looked over at me with a deeply skeptical eye, as if to say, "I don't trust you one single bit."

I bent down and, as gently as I could, poured the warm water around Chester's body. The ice hardly melted. I'd need a lot more water. Finally, after much water and a little melting, Chester pushed himself up with his one good leg and hobbled off, leaving a bunch of tail feathers sticking out of the ice. Those feathers stayed there for the next two weeks because the temperature never rose above freezing.

This episode weakened Chester considerably. He was less able to keep up with the others, and his tail feathers never grew back. In early spring, he was gone, probably caught by a coyote or a fisher sneaking in from the edge of the nearby forest.

The other two stayed with us for a few more months, but then both disappeared within a week of each other. Soon after, we noticed a scattering of green and gray feathers near the garden shed.

It's no coincidence that Rachel Carlson named her book *Silent Spring*. She knew that it would be the absence of birdsong that we would notice first—and miss the most. So it was with the duckness. They'd been such noisy companions and so fiercely determined to be heard that the disappearance of their voices left a small soundless space in our lives.

Pedro and the Chess Game

Pedro was a stunning buck, weighing more than two hundred pounds and with a majestic set of curled horns. He had black and silver hair, an expansive chest, a swagger in his walk, and a pendulous pair of enormous testicles. Despite the twenty older receptive does in his own pasture, Pedro had become obsessed with trying to smash his way into the adjoining field where a much younger buck was courting our younger does. In the process of hammering away at the wooden fence, he became tangled in the off-set electrical rope.

Envy, or perhaps lust, doomed Pedro. But we weren't blameless. We hadn't seen clearly enough what could happen. We hadn't made it safe for him.

Death came often to our farm, and I began to think of it as a presence that always hovered nearby. In *The Seventh Seal*, Bergman tells about a knight who, returning from the Crusades to find his homeland gripped by the plague, meets Death himself and challenges him to a game of chess. The knight, played by Max von Sydow, is weary from a long war. His face is gray, his eyes vigilant, his gestures deliberate. The outcome of the game is not in question, but the sly knight buys a bit of time for himself and his companions, distracting Death long enough so they can have a few more moments of yearning, silliness, and love.

After I stood next to Pedro long enough to understand what had happened, I turned off the electricity to the fence, unraveled the rope from his horns, and pulled him a short distance into the pasture. I worked deliberately with a weary sadness in my heart. Pedro lay in the grass on his side, a blue eye gazing up to a cloudless sky. I had failed him. Death had come to the farm once again, and I wasn't there to suggest a game of chess.

January 11, the Fire

Priscilla

Why was that damn dog barking?

The ruckus came from just outside my bedroom window. It didn't sound like Butch was chasing deer away from the house. No, it sounded like he was spinning in circles and barking insanely. Why would he do that? It had to be past midnight.

I sat up and looked to the other side of the bed. The blankets were flat. That's right; Henry was staying in the city.

Alone, I jumped out of bed with a sudden sense of urgency and scrambled toward the noise.

Why was the night sky pink?

And what was wrong with that damn dog?

The night sky was much too bright. Red and orange and yellow flashed high in the air behind the big barn.

Flames! The goat barn was on fire!

In shock, I froze at the window, unable to move from the horrible sight. Surely, this was a bad dream. I'd wake up.

But I didn't. I could only stand there and stare.

Pris, think. Think!

First, call 911 for the fire department.

Second, get on warm work clothes, fast. The temperature

was fifteen degrees when I went to bed and it was colder now. Toe warmers. Hand warmers. Boots. Waterproof gloves.

Third, keep your mind on the job. No breaking down. Not now.

Bundled up in minutes, I ran toward the roaring fire as it consumed my goat barn and everything in it. The closer I got, the more I shifted into automatic pilot. *Get water. Will the hoses work? Check if anything is alive.*

Coming closer, the flames began to scorch my face. *Nothing could be alive inside.*

How long until the fire department arrives?

A dog still barked frantically from the back of the burning building. The back paddock! I rushed around the barn. Arnost had gathered his herd of does at the back fence, as far from the fire as he could get them. He was manic, already pushing his does toward the gate as I hurried to open it. Without even looking my way, he chased them into the middle of the big field and away from danger. Barking wildly, he circled them; any goat out of line suffered Arnost's crazed wrath.

My mind kept screaming, *Do something.* But what? *Something to save the lower barn. Stop the fire from eating anything else.*

Despite the cold, the water hose worked. I sprayed water on the ground around the burning barn, and back and forth along the fence. I soaked everything I could, running from one place to another, thankful I'd remembered my waterproof gloves.

My chest tightened with pain from gulping the frigid smoke-filled air. I coughed—or cried—not sure which. The only smell was fire—a horribly hot smell that blistered my brain with panic. Tears and snot mixed with ash in my mouth. I gasped for breath—or maybe sobbed in pain and fear.

Exhausted, I stood in front of the fire and listened for the fire trucks. My mind was mush. I could only wait there and hope for the noise and lights for help.

Yet, I couldn't help but note that beauty surrounded me. The flames reflected on the snow, creating a unique light show—one of stillness and peace. The spray of water steamed in the wind before it turned to ice. Splendor and art amid this terror, loss, and sadness. Those moments were the only quiet ones I would have for days.

Finally, sirens carried through the quiet darkness—distant, then closer, then fading away, then coming back. An eerie, otherworldly sound. My body was numb, but I managed to wave the hose in the air as if the spray of water was made of lightning bolts and could lead the firefighters to me.

The livestock in the other barn down the hill were far enough away not to be affected by the fire. Just to be safe, the dogs had chased all the animals out of that barn and into the adjacent field. They were huddled at the back of the field, as far as they could get from the trauma and smells of death and destruction.

The fire raged. Surely, fire trucks would be here soon.

Then I realized that one of my favorite things about our farm was a big problem: it was not visible from the road. The firemen couldn't find me. They drove up and down the hollow road, missing our driveway again and again. I could hear the sirens coming and then passing. Finally, they found our small dirt road through the woods and barreled toward our barn.

The firefighters were on top of the situation in minutes. I stood, still clutching my hose, while they quickly took their stations. Even though they were volunteers, they worked together as an experienced team.

"Oh, shit! There're dead animals in here!" yelled a firefighter.

My heart stopped for a minute or two. I dropped down and sat on the top of a burned-out bucket, trying to understand what was happening, what had already happened. My hands still gripped the hose. The big waterproof gloves that had kept my fingers from freezing were now frozen to the hose. I couldn't get my hands out. My head dropped between my legs. *Breathe, Pris. Just breathe.* Somehow, I managed not to faint or vomit. Just cry.

Tears trickled down my frozen face. Cate, an old-line Savanna doe, and her doeling, Sweedey, were my favorites. I had put them in the barn right before I went in for the night. I didn't usually do that, but I felt sorry for them. I didn't want those special animals to be outside in the bitter cold. When I closed the barn door, they were warming in the glow of a heat lamp. That was my last sight of all the does and their small kids.

I had bred six of my best Savanna does to get birthing timed for the kids to be sold to the boys and girls in the 4H group at the local high school. New market, new money for a struggling little business. My plan. The weather turned dangerously cold, and babies had to be kept out of such harsh temperatures. I locked them in the barn with heat lamps.

I lost eight very important does and their offspring. The scene played over and over in my mind.

"Mrs. Ireys, do you have anyone here with you?" a fireman asked in a kind voice. He bent down and talked to me as I sat on the bucket.

"No," I answered.

He asked me more questions, but I couldn't hear him. His mouth was moving, but I didn't know what he was saying. There was too much noise in my head.

After the fire was out, some of the firefighters coaxed a group of goats and dogs down to the lower barn and fixed what they could of the burned, broken fencing. They stacked the metal gates that weren't ruined neatly against a safe fence—an act of extraordinary kindness that I didn't notice until the next day.

The guys also knew that, in this terrible cold, the animals needed water. They checked my remaining water troughs and assured me that the troughs were full and the heaters working.

Still sitting on the bucket, I thanked them. At least, I hope I did.

Arnost still had a group of does held hostage in the middle field, circling, nipping, barking to keep them close. The wind had grown fiercer, and the temperature must have been in the single digits. I told one of the firemen I had to get those animals in, too. He nodded and helped me off the bucket.

"Go on in, Mrs. Ireys," he said. "We'll take care of it."

"No, thank you. I need this bit of normality." I hobbled away to retrieve Arnost and his herd. Behind me, the firefighters began putting away their gear.

Walking into the dark, away from what had happened, I felt the shock of a brisk, frigid wind. I kept waiting to wake up from this disaster. How could my body be so numb? *Keep walking. Get the goats in. Keep going, Pris.*

Arnost was happy to see me, and his does were happy to run to the only barn left and join the rest of the goats. Even in the dead of that cold night, smoke still circled up and disappeared into the darkness. Red embers glowed amid the soaked ashes of the fire.

I don't remember if I went into the house before all the fire-fighting folks left. I do remember seeing the long silhouettes of trucks maneuvering through the bottom field, heading back down our dirt driveway. Their trucks' yellow, red, and white lights snaked away, turned onto the hollow road, and disappeared into the darkness.

Back in the house, I felt strange and very lonely. This was no normal night. This was strange in a bad way. Henry and I had talked at length about the problems of living separately for two or three days each week so he could stay in Washington, DC, and work in his office. We had agreed this arrangement would work to meet most of our needs. But I had lied. This arrangement did not meet most of my needs, and the night of the fire showed why. I was alone and needed someone to be with me, to hold me, to share the enormity of this shattering event, to comfort me. But no one was there when I smelt of death and fire.

No one was there and our plan wasn't working. I couldn't remember how long I had been living in this arrangement. I was terrified.

The room was silent. I stripped off my clothes and left them on the floor. I wrapped a robe and a heavy sweater tightly around my body.

Sitting in the living room with a small glass of scotch, I still smelled the combination of fire stink and wet grime on my hair. I couldn't get rid of the terrible taste in my mouth, so I slugged another scotch. My eyes still stung. My hair was still matted. All of it kept reminding me of the reality of what had just happened.

I think I dozed off for a bit. Eventually, I found myself standing in the shower, letting the hot water wash away the grime and warm my shivering body. I looked down at the filthy water coming off my body and scrubbed myself from my hair to my toes, my tears mingling with the water coming from above. I had not called anyone. Not that I didn't have friends who would be here for me. I was too numb. There wasn't much of me that was left to think right now. I drank a large glass of water.

I finally called Henry around two a.m. The call was short and emotional. The first thing he yelled was that he had not changed the insurance. He said he'd be home in the morning as early as he could.

I still don't remember where I slept that night.

When I woke, I started to think again in strictly necessary tasks:

First, make coffee.

Second, eat something.

Third, make some phone calls.

Fourth, feed the livestock.

After breakfast, I climbed into the truck and drove up the hill to start the painful job of sifting through the ashes. Henry drove in. We spoke to each other from our vehicles. He said he had to take a call and we'd talk later.

My mind and body went back on autopilot. I pulled up to the remains of the barn, only a charred post, melted equipment, and bodies of dead goats. The smell of death hung heavily in the thick air. I needed to get the corpses out of the barn in a hurry. The bodies of the does were bloated and grotesque, wrapped in tight charcoal-coated skins. Thankfully, I did not have to deal with the kids' bodies—they had been incinerated.

I cried as I worked that morning. Being in the middle

of the soaked, burned-out building leaning on my rake, I was overwhelmed, alone with this huge mess. Then, I heard the hum of Eddy's red truck. Eddy had helped us on the farm for years and seemed to show up when I most needed him. He'd done it again.

He and I got to work, never mentioning the fire or the deaths. We loaded the bloated bodies in the back of the truck and took them to a dumping spot far down the hollow, on the other side of the mountain. A small comfort was knowing I wouldn't see or hear them being eaten by the predators.

Eddy allowed me a few minutes to cry and stand in silence, then he gently put his hand on my shoulder. We needed to get on with the rest of the work.

The clean-up days were above freezing, which was good, except the thaw of ice and frozen ground brought deep mud. We moved bent, burned skeletons of feeding troughs, light fixtures, and hay racks across black cinders.

"We've got to put that stuff in the truck and get it out of here now. Are you ready?" I asked Eddy as I blew my nose for the hundredth time. "I need to stop smelling the death stench," I whispered to him. He understood and we started driving.

On the way to the dump, my mind kept tallying the business losses: Did we have enough feed to get through the week? It was January, and we'd just lost a lot of our hay. This changed our marketing plan for the next year completely. We lost six of our most important Savannas does. We needed more sheds for the kidding season. We lost the goat barn. What was I supposed to do first?

And then my mind would jump back to the emotional stuff. No matter what I thought, it was sad. I had no plan for how to get through this. I felt so alone, even though Henry tried to be on the farm as much as he could. Was he ever really there? Deep pain throbbed in my muscles, so I tried to focus on the discomfort in my hands and back, not in my brain and heart. Fires are often about mistakes. This one was no different. The next couple of days were bookkeeping marathons and screaming matches about mistakes that both Henry and I had made. I took many long, tearful showers. Calls came from family and friends showing support.

I found myself looking out the windows, checking the barns throughout the night, every night. I paced silently though the house, dabbing at my tears. The reflections on the walls made by the fire in the wood stove terrified me. On some nights, they still do.

But, as always, good mingled with bad. We were the recipients of gifts—bales of hay and sacks of grain from farmer friends. My grain supplier gave us free bags of feed and minerals. Our neighbors and friends and family sent food. Lots of food. Was I going to eat my way past this sadness?

In February, our neighbors up the road began to plan a barn-building party for the early spring. A week later, a goat breeder friend called me with a business offer that helped us stay afloat.

Winter turned to spring and the weekend for our barn-raising arrived. Family, friends, and neighbors showed up with tools, boots, and determination. A few brought their musical instruments.

Henry had ordered the lumber, bought the nails, and generally organized the project. Folks started arriving the day before. In the country it is wise to plan on unexpected stuff and people. They came with equipment, trucks, and tools. They got their working orders from Henry. They camped in tents, campers, and the big barn. They worked together in different groups; some working on the walls, some on the roof, some on the interior doors. Everyone was thrilled when the frame was finished on Saturday afternoon. Afterward, the group walked to the house for roast pork, apple pie, and some old-time mountain music. Everyone promised to be back the next day.

And, yes, they came back the next morning and worked most of the day. The barn was built in a weekend, thanks to the generosity of fellow farmers, new and old friends, and our children. Henry's careful organization helped pull it off. The whole effort soothed my still-raw wounds. I had a new barn, and my goats would have a new place for the kidding season.

The crew left as the sun set on a warm evening, their car lights disappearing down our dirt road—different lights from those I'd watched on the cold morning of January 11.

About a month later, our dear Patty, a big Spanish doe, was the first to birth in the new barn. Triplets! All nice size and healthy. Arnost made his dog cave in the barn's back left corner; as the new kids came along, they crawled all over him. In the warmth of the early summer sun, dogs, does, and kids took their naps in the shade of the new barn.

As much as my memories of that night sometimes dragged me back to a deep, dark place, I needed to stay strong for the next generation of goats, cows, pigs, and ideas. A new season was in the air; our friends and family had helped us out more than I could have imagined; Henry decided to work from home full-time. Later, he'd describe our marriage's survival as a "miracle, miracle." I would describe it as a leap of faith. We live on the farm together and make most decisions together.

Together, not alone.

January 12, the Fire

Henry

My cell phone rang about 2:30 a.m., waking me from a deep sleep. I saw the number. *Oh, God. Bad news.*

"Hey, what's up?" I asked, still groggy.

"The barn's gone." Priscilla said. "It burned. The goats couldn't get out."

"The barn burned?" I repeated, struggling to understand the meaning of her words, grasping at their implications.

"That's what I said."

"Oh God, we don't have insurance. What happened?"

"Do you have hearing problems? It's gone. There's nothing left."

"Okay," I said, now fully awake, aware now that her voice was numb, that she was barely holding on. "I'll be there as soon as I can. The first train doesn't leave until, I don't know, maybe 6:30." I sensed then that I might never know what really happened.

"I gotta go." She hung up. I called back. She didn't pick up.

I tried to imagine the barn gone, even though it hadn't been there long. We'd needed more barn space after we'd expanded the herd last summer. We finished it in mid-September; so it'd been there maybe four months. A pole barn, twenty by thirty feet, with a sloping tin roof. Did the tin melt? Was the barn really gone?

And what about the goats? How many were in there? Priscilla usually kept the Savanna goats in that barn because they were special to her. Were they all gone? That would be horrible.

I climbed out of bed, dressed, and arrived at the train terminal long before the 6:30 commuter train left. It pulled into the last station on the line around 7:45, where I'd parked my car. Shortly before 9:00, I skidded to a stop at the bottom of the dirt road leading to the farm. Our red truck was coming down the hill. Priscilla was driving. Eddy was in the passenger's seat. I rolled down my window.

"What are you doing?" Priscilla asked. "I thought you were going to be at work today."

"Pris, come on." I replied. "I told you I'd be home. Where are you going?"

"We have bodies in the back. We're dumping them over the hill on Spring Gap. We'll see you later." She drove off. Her face had a distant, absent-minded look. She was hanging on by a thread.

I drove to the house, thinking about how hard this was going to be for her. She loved those goats. She'd tell me about them— each of them with their particular habits and personal quirks:

You know, Henry, Craig is really unusual in his enjoyment of the human touch. He's my lap goat! Even at three months, I can see that he's going to be a magnificent buck. He has that big chest and that herd-sire swagger. He'll keep his kindness, too. I can tell. He's already pushing bigger goats around, but only when they are mean to him. He'll come walking over to me with those curious eyes and stretch his head out so I can scratch between his horns. He loves his humans' hands.

I walked up the hill to the soggy, charred remains of the barn—a patch of charcoaled cinders, gleaming black and grey in the bright morning sunlight. Sections of the tin roof were bent into weird shapes. Hundreds of nails, a few still shiny, lay amid the rubble. I found a hinge whose two sides were fused together and twisted, a small, blackened sculpture of a tornado.

I had to turn away and look out over the pastures, dotted with patches of snow. In the rubble behind me lay the withered ruins of many plans, plans for a distinctive Savanna herd, plans for a strong line of Savanna-Spanish crosses, plans for special goats. Priscilla needed these kinds of plans. She'd have to make new plans. I didn't know how to help. All I knew for certain were her stories:

Craig's mother, Lydia, is stunning. She has perfect confirmation. Perfect. Straight back. Classic Savannah face, an elegance born from many generations of good breeding. Legs with beautiful lines. She walks more gracefully than any doe we've ever had. I can see her as a model on a runway. A graceful sway, a hint of flounce, but mostly just perfect proportions strutting along. And she knows it. Not that she's mean, but her elegance makes all the other does feel shabby. So they keep their distance. Except her friend Ernestine, who more or less defines average. Funny how they wait for each other so they can both sashay out to the pasture together. They're the best of best friends.

One wooden fence post survived the fire. It still stood, about five feet from where the barn's corner had been. Charred on the side that had faced the fire, it was still standing because Priscilla must have been able to hose it down. It was a lone witness to the night's carnage. I took my hand from my glove and ran it along the post's brittle blackened face, feeling its rough,

Braille-like surface. Its corrugations told the story: inside, a first dim crackling, outside, dogs' growls turning to shrill yelps; inside, goats' fear-filled screams; the flames' furious roars. Lives had vanished.

Two days later the insurance company confirmed to us what we had suspected: the barn was not covered under our homeowner's policy. The next day a six-inch snow finally cooled the embers. We sat at the dinner table, with plates of half-eaten brisket and an empty bottle of red wine. All the difficult parts of farming and all the difficult parts of marriage mixed themselves into one huge emotional stew. Anger morphing into rage, frustration turning into blistering blame, grief making speech impossible, accusations hurled like thunderbolts, an adrenaline-fueled exhaustion that banished sleep. It was all there, and it wasn't clear for many weeks that the farm would survive, that we would survive, that we would begin to make plans again.

But every death, even a dream's death, opens a space for others. A fire's fury fades. Time blunts the razor's edge of agony.

We did begin to plan again, thanks maybe to some stubbornness and some voices inside each of us, and certainly to an outpouring of support from friends, who made meals for us and sent gift boxes and offered to help us rebuild. Later in the spring, we had a barn-raising with a lot of help from friends—plus fabulous food and impromptu music. We'd been given the chance to heal through simple kindness. We, and the farm, would never be the same.

Have I told you how wonderful Cate's daughter, Sweedey, is? Myles is her father, so Sweedey is Savanna royalty. She's a spitfire, a little Miss Everything. She gets up on the rocks and jumps off with that funny side kick that you see in all those stupid goat videos on YouTube. She'll do it a couple of times and then go running over to her best friend. "Wasn't I great! Come on. Let's climb the big rocks!" I love her to pieces.

The Forgiving Land

Henry

The battered sixty-foot logging truck backfired its way down the steep dirt road behind our house. Burdened with tree trunks stripped of their branches, it bumped across muddy potholes, swayed along the ledge above the creek, and lumbered finally onto the blacktop. The driver—a heavyset man with a thick beard, a cigarette in his mouth, and a history of heart attacks—shifted gears and headed over the mountain to the sawmill. The staccato eruptions of the truck's ancient muffler grew distant. I waited, resigned to the fact that a second truck would follow soon.

The trucks made this trip about a hundred times from July to September that year, hauling their harvest from twenty acres of our land. The loggers cut everything but three clumps of trees. After the crew had gone, logging roads crisscrossed the bare hills, stumps dotted the denuded slopes, and treetops with lifeless leaves lay scattered across the landscape. Like a bomb blast, the logging had left a casual, indifferent chaos—random piles of dirt, cockeyed heaps of tree limbs, and big trunks lying dented and half-smashed by the skidder. The few remaining trees stood desolate and haunting, marooned in a sea of tangled branches. I looked at the ugly naked land I loved and wept.

Priscilla and I had caused this devastation with an idea that went wrong. Instead of remaining passive witnesses to the consequences of two hundred years of careless timbering—which had left a sparse forest with few species of trees and limited understory—we wanted to be active stewards of the land. Our idea was to clear limited patches of forest to create islands of sunlight for new growth. A simple idea that would, we hoped, bring a diversity of grasses, bushes, and young trees into the forest.

We hired an experienced forester to help. A lean man with gray hair and a pleasant smile, he spoke at length about working with landowners and their forests on many previous jobs. After several treks together through our woods and discussions of the lay of the land, he came back with a map showing the areas to be cut with circles indicating areas to be left untouched.

In retrospect, I wonder how we missed something so obvious. On the map, the proportion that would remain untouched seemed small relative to areas that would be cut. We had agreed that circles of trees would be cut within the forest to create islands of sunlight scattered through the woods. But the map showed circles of trees standing within a clear cut. I didn't see the error.

Like the Titanic's wireless operator who ignored a warning about icebergs because he was busy sending messages for the passengers, I was distracted by the demands of my day job. I missed the danger signals. Also, I trusted that the forester understood what we wanted to do because he had said it was a good idea. I didn't realize he had his own ideas and his own rules.

"Are you sure those circles of trees are big enough?" Priscilla asked at one point.

"This is the way to open up the land," the forester explained, tossing the map on the kitchen table. He stared at us as though we were idiots. "I can't get a logger in here if they aren't going to make some money. It's not worth it to them to take a few trees here and there. This is the only way to do it."

"Are the boundaries marked clearly enough?" Priscilla's brow wrinkled in concern. The forester had spray-painted orange swatches on a circle of trees but faced the marks to the inside, where the loggers could not see them easily. And the outside boundaries of the clear cut were poorly defined. The ambiguous markings made us worry the loggers would cut more than they should. Of course, this was the first time we had logged our land. What did we know?

"I've worked with all the loggers around here. They know me, they know how I mark," the forester said, annoyed she would question him. "They won't have a problem. Don't worry."

A faint doubt rippled through my thoughts like a distant alarm announcing trouble. Once again, I ignored the warning signal.

The promise of making our vision come alive on our land— of seeing small pastures of red clover and purple pye weed— pushed us onward. We signed a contract that specified, among other conditions, the percentage of the timber sale the forester would receive. He sent a technical description of the job to nearby logging companies, asking for bids on payment per ton for softwoods (mostly pines of varying kinds) and per foot for hardwoods (mostly oak, maple, and hickory). What we didn't realize early enough, and should have, is the obvious: the more trees the loggers cut, the more money they'd make from sales to the sawmill and the larger the profit going into the forester's pocket. A simple system with incentives to maximize the cutting.

Before long, the loggers were ignoring the boundaries, grabbing big trees that lay beyond the marks. The forester rarely stopped by and provided little oversight. We finally understood that our original idea for selective cutting had disappeared like the trees themselves. Priscilla walked the hills with orange tape to make the remaining lines vividly obvious.

"You are not taking this tree," Priscilla announced one morning, when the jaws of the big saw were opening wide to cut a magnificent white pine that had shaded us and our horses many times—and that should have been off-limits. She stood in front of the tree and did not move. The loggers turned toward another part of the forest.

"Stop cutting," we finally told the owner of the logging company and the forester. We re-negotiated our contract with the loggers and paid the forester the percentage of what was sold, despite his claim that we owed him the percentage of the total expected sale.

But the damage was done. During our first decade on the farm, I'd become intimate with those woods. I knew the lean of certain trees, the twist of particular paths, and the maze of branches in a small stand of mountain laurel. And now? No canopied path, no shadowed trail, no whiff of damp earth even on a summer afternoon, no dappled rocks, no stand of mountain laurel. I had lost a community of friends. A soulless void had replaced the material body of the forest.

How could I have so hurt what I had so treasured? "I'm sorry," I whispered many times as I walked through the devastation. "Can you forgive me?"

Guilt and anger traded places often. I was angry at Priscilla because it had been her idea in the first place to meddle with the forest. I thought she'd just take on the responsibility of overseeing the job. Why, I kept thinking then, was she always needing me?

I was angry at the forester because of his greed. He must have known he was ignoring our intentions. I can find no other explanation for his actions other than self-interest.

I was angry mostly at myself for ignoring my instincts and listening too late to Priscilla's alarm. During my walks through the devastation, a maelstrom of emotions—shock, rage, remorse— swirled so intensely that sometimes I could barely keep going. And yet, the denuded hillside kept pulling me back, as if it had something to tell me.

One October morning, I stood at the clear cut's highest point and watched sunlight play upon the corrugated surface of a large stump. Below me, the undulating curves of the newly naked land struck me as sensual, seductive. The hills across the valley, not visible when the trees still stood, were speckled by the outbreak of autumn.

A few weeks later, a soft rain soaked the exposed dirt. A day later, as she walked through the clear cut, Priscilla noticed a few green sprouts. New grass? Sweet goldenrod? The poison ivy favored by our goats? Re-birth had signaled its inevitability.

Living in one place for a long time is akin to a long marriage: inevitably, unexpected events reveal the fault lines in our characters: calamitous lapses in loving, hollow self-centeredness, misjudgments about one's supposed gifts, sour stubbornness beneath saccharine affability. I'm probably no more flawed than anyone else—up close, we're all a bit raggedy—but it was hard to see my shortcomings so obviously revealed through the land itself. My failure to understand the timbering contract led to a desecrated landscape. And since then, I've done little to stem invasive species, I've ignored pastures that need lime and fertilizer, and I've yet to develop a coherent "management plan" for our forests. Maybe the best plan is to have none.

Many years ago, I asked the land whether it could forgive me. Now it's crowded with Virginia pine, thickets of multiflora rose, wild blackberries, and young hardwoods growing from old stumps. Deer—along with rabbit and fox—bed in its dense thickets. Bears feast on summer's blackberries. Cedar waxwings swoop around, chatting to each other. Perhaps all this growth and commotion is the land's way of answering, "Yes."

The Old Oak

Priscilla

Standing at the edge of what used to be the back meadow, I stared at the sixty-foot tree surrounded by multiflora rose bushes and briars so thick I couldn't get closer than twenty yards. "He's just slow to leaf out," I said to myself. But I had to admit the other trees were already in leaves.

The oak was an old, wounded warrior, dying but still standing. There was a dignity about him. He clung to his branches and a few leaves, like a protective father. As I watched, a leaf flittered to the ground and vanished into the multiflora roses.

Gretel Ehrlich, writing about remote regions in the American midwest, describes the lessons that impermanence taught her. "Loss," she writes, "constitutes an odd kind of fullness; despair empties out into an unquenchable appetite for life."

Did I not see this coming because I was too hungry for a certain kind of success?

Years ago, when I still chased dreams of making our part of the mountain better for our goat herd and cattle, we logged twenty acres of our land to make more pasture and good places for hogs to roam. It went badly. The timbering hurt both the land and my soul forever.

What I had wanted—selective logging to create savannas bordered by trees for wind brakes—turned into a disastrous clear cut, taking almost everything and leaving a mess. The hardwoods we intended to bring back to our farm would not be planted. The pastures would not be created. No hogs would roam the woods.

I had put my trust in a forester, had given him a drawing of what I wanted to do, had walked our land with him, had watched him mark trees, had talked about my plans over and over. What I did not take into consideration was the fact the forester and loggers make money on the trees they take down.

My trust in so many things was in question. As I looked out over the destroyed acres, I realized that I had no plan B for this screw-up.

I'd talked to both the forester and Henry about the obvious miscommunication. I saw the work was not going the way I had planned. I yelled warnings. Unfortunately, Henry was too preoccupied with his day job to heed my hollering. He had put his time in on this project and I was to take care of things and not bother him anymore.

"But isn't that in the deal! We make all the decisions together." I yelled to myself and wished I were yelling at Henry. This farm is a full-time job, and I loved it. We both loved it. It was our chance to do something meaningful and I wanted to farm with whatever energy was left in my stinking, fragile body. Didn't he want that too? I knew he loved this land as much or more than I did. It was pertinent to his well-being.

Finally, on the afternoon I got him to stand and stare at what had been done to the land, I saw in him the sadness that takes your breath away. He saw what I had been yelling about.

Finally, after a terrible argument that night, we told the loggers to stop cutting. They left not long after that, but enough damage had been done by then. Our lack of attention had destroyed the land. Nature would have to lick its wounds for years.

There was no cleanup of the devastated ground, so multiflora rose took over.

Now I'm looking at the tree I rushed through the war zone to save from the loggers years ago. The oak was the only thing left in the back meadow, and I couldn't bear to see it cut down. I remember leaving the house in a tearful rage after pleading with Henry to come out and stop the logging. "I need you!" I screamed. When I saw the machines moving toward the old oak it was all I could do to stop another tragedy.

This logging catastrophe took almost every tree on twenty acres. I was determined to try to save a few.

I was not surprised the old oak was dying. He couldn't live any longer within this broken world. His community was taken away in broad daylight. No one heard his calls for help. No one heard my calls. Greedy dreams and powerful machines took it all away.

The distress calls must have gone out when the cutting began. The wildlife had seen this sort of murder before, and they fled. But the trees could not move. We know trees communicate with each other. At the time of the logging, did they wire goodbyes to their neighbors?

Saved from the logging fiasco, the old oak stood alone in a war zone with few survivors. He must have heard only sorrow coming though the mycorrhizal network. He had dodged the carnage but was alone in the meadow.

As the loggers worked, I heard the trees crash to the ground, scraping against each other as they fell. The sound was a sad hymn of their woes. The loggers said it was just settling wood. "It takes a few days for the collateral damage to stop falling," they said. "That's what you're hearing, Mrs. Ireys."

Collateral damage. The trees and branches that were left to rot were collateral damage. Isn't that what they call civilian deaths in war?

When the loggers had left and it was very still, I'd sit on a stump and listen to the sad hymn travel through the mountains. The forest had lost some of its own. These ancient hills had seen this carnage many times for hundreds of years. The horror was new to my eyes, and I wept.

A tree that took a hundred years to grow took the loggers about twenty minutes to cut down. Another ten minutes to strip the tree of all its branches. Another ten minutes to load on a truck. Those branches were left on the ground to rot. If nothing more is done to the land, the rotting wood will be good soil in a few hundred years or so. But what chance does land in the eastern part of our country have to be left alone for more than a hundred years?

I looked over the five-foot wall of multiflora rose, scrub pines, and bushes growing on the wasted land between me and the dying oak.

After the hard winter and the oak's other wounds, including from a lightning strike, it was too much for him to survive. The once massive, proud tree just couldn't stand up to the stress of life anymore. He couldn't listen to any more sad tales from his few neighbors or their pleas for protection. He had given all he had and was sick.

I wanted to go to the oak and wrap my arms around his trunk, but I would've had to cut through the rose thickets with the tractor. For what? A closer look at death? I knew what death looked like already.

The everyday decisions I needed to make kept me going. Henry had gone back to the city and now the aloneness was my comfort. I didn't have to pretend and act happy or nice. I could do my work and let this place heal my wounds and my guilt. The place, my keep, was my saving grace again.

It wasn't all sad, I reminded myself. The old warrior had had his glory days.

The meadow had once been filled with goats and dogs. Kids frolicked around the big oak tree, bit off autumn olive leaves from the many bushes, sparred with one another, took naps as they lay around his big roots.

He had been home to many critters. I saw my first fisher hanging from one of his branches. It took forever to get the dogs away from the bottom of the tree so the fisher could run for the safety of the mountain.

The huge solitary oak was the main attraction of the back meadow. The other trees were dwarfed by the spread of leaves, branches, and the height of its impressive trunk.

He was protection for the herds when they were caught in storms in the back pasture, and the home for many nesting birds as they passed through. Nick, our son, built his first deer stand around the oak's trunk and saw his first bear strolling under its branches.

Before the farm, wild turkeys and grouse wandered in the fields. A beautiful pair of red fox lived in an overgrown gully in the middle of the back field. The years of livestock protected by guardian dogs drove all the wildlife out of the farm. Now some of these critters are coming back.

Where they had been a nuisance before, the animals now are rewilding the fields. If an eagle flies overhead, I hope it gets one of those rabbits eating in my garden.

With the herds gone, we only need enough pasture and hay for two horses.

I find myself smiling. No, it is not all bad, and there is good in what we still have.

I walk across a run bed to another part of the clear cut to find comfort in the sight of the only surviving giant white pine. We figure she's nearly eighty years old.

I had saved her, too, from the loggers. Underneath her broad spread of limbs lay a soft layer of pine needles and a dozen of her children: saplings, a few feet tall, holding out their spindly arms. Henry and one of his friends later transplanted a half dozen of these small white pines to a spot above the pond. They will be magnificent someday. Maybe our grandchildren will appreciate them for their dignity.

I stretch my arms around the old white pine. Her trunk is wider than my arms with fingers pulled as far as they can reach.

I feel life in the old tree.

I close my eyes and hug her as hard as I can. Her astringent scent fills my nose.

So much life.

Gifts

Winter's Wood

Henry

A December evening. Deep darkness presses against my living room windows. The temperature's dropping: forty degrees as I write this, but I'll wager freezing by midnight. Hulking clouds that lumbered over the mountain at sunset will bring snow before long.

My woodstove whistles to the wind's sudden gusting. Like small epiphanies, a few embers flare fiercely behind the stove's front window, itself framed by three cathedral-like iron arches. I slip in another log and it catches quickly.

Our living room is snug and warm because I harvest, split, and stack wood from dead trees found in the forest. Before autumn's first sleet, I've filled our woodshed with quartered logs and small rounds of dried locust, elm, oak, walnut, sassafras, cherry, and ash. With winter's advent, the logs are stacked in our living room wood chest, within easy reach of our stove. I have led them on a journey from our land's forest to our stove's firebox—a journey of considerable labor (much of it welcomed) and ecological ambiguity (most of it unresolved).

I began building this winter's woodpile six months ago on a beautiful June afternoon when I drove my tractor toward the forest behind the back pasture. I was looking for standing dead trees or for trunks blown down during spring storms. A couple of medium-sized dead locust trees stood close to a flat dirt road easily reached with the tractor. It would be simple work to cut them down, chain them together, and drag them to the log splitter—a good start on the four cords we needed to keep our house warm during winter.

Then, I remembered that a much larger oak—long-dead but still solid—lay at the far corner of our property. It had recently fallen above the back pasture's electric fence, near the road that leads to the mountain behind us. The hill was steep over there. Pulling it out would be tricky.

"We can get those smaller trees anytime," my younger self said to me as we sat on the idling tractor. "Why not get the big one now?"

I nodded in agreement.

And then my discomfort began. Unlike my younger self, I knew that removing dead trees also removes a host of residential opportunities for beetles and salamanders, mice and rabbits, worms and mushrooms—living creatures and plants essential to a vibrant forest ecosystem. Taking that big oak from the land means fewer homes for ground-nesting birds, less shelter for various rodents, and no morels for future mushroom hunters.

But, of course, it's not a simple matter. The less wood I use to heat my home, the more I depend on propane, a fossil fuel that requires a lot of energy to produce and that adds carbon dioxide to the atmosphere as I burn it. Even though I have a highly efficient woodstove, our contemporary propane stove is even more efficient—at least, according to several well-documented articles I found on the internet. But propane is considerably more costly, even when I factor in the gas I use for my tractor and chainsaw. Cost matters. So does the forest environment.

I sat on my tractor, feeling the sun's warmth. Too bad that humans have yet to reliably and cheaply duplicate the fusion that powers the sun. Someday, maybe, my dilemma won't be so acute. But there was no magic solution at hand. I needed heat for the winter, one way or the other, and I was fortunate to have a choice about how we'd stay warm. Should I turn the tractor around and rely only on propane? Or keep going and harvest the wood?

My younger self squirmed. "Hey, this is crazy. *It's just a few trees.* Trees are down all over the forest. Some trees die every year. We don't use all those trees. We'll never use them all. Right? There'll still be plenty of homes for those creatures you're worried about. And you can donate the money you're saving to a land trust. It can be your own version of carbon credits. How about that?"

"But that's not the point," I replied. "The world is in trouble because everyone thinks they're just taking a few trees, that the damage they're doing is no big deal. That's not the way to go."

"Hey, at least you're being mindful about this. You're not just blundering forward. Give yourself some credit."

"That mindfulness spiel is poppycock, and you know it." I shook my head slightly, trying to dislodge my irritating companion. "Lots of folk say they know the trade-offs when it comes to difficult decisions. They hem and haw to show they're thoughtful and conscientious. Then, they do what they wanted to do in the first place and say they're being mindful. They're the worst."

A red-winged blackbird called out from the nearby tree line as if she were contributing her opinion. I couldn't understand her language and maybe she wasn't talking about me at all. Still, I'd like to have known her vote.

In the absence of a clear best choice for how to heat my house, I remained quiet for a few minutes, feeling the vibrations of my rumbling tractor.

And then a simple emotion announced itself: I love sitting in my living room during winter months and watching the stove's pirouetting flames and cat-eye embers. Depending on the evening and the weather, it's a good time for desultory conversation, for the joy of affectionate companionship, and even, sometimes, for the quiet torture of writing. Does the emotional power of a stove's fire figure into this dilemma at all?

I made my decision, shifted my tractor into low gear, and headed for the big tree.

The hill was steeper than I had remembered, and the tree thicker, and it lay closer to the pasture's electric fence. The details of what happened next are unimportant. Let's just say that a dead oak, twenty inches wide at its base and forty feet long, can pick up more speed rolling downhill than a man—thirty-four inches at his waist and an inch shy of six feet—can handle. My good idea flattened several plastic posts holding the electric fence, and nearly leveled me. One more close call for the record.

I eventually chained the trunk to my tractor and pulled it into a clearing. In a few hours, I'd cut it into foot-and-a-half rounds, load them into the tractor's bucket, and dump them near the log splitter. I returned to the flattened fence and climbed down from the tractor to reset the posts, forgetting—of course—the electric rope was still hot. My involuntary response was immediate and loud, but momentary.

"Nice one," my younger self sneered as I trudged back to the barn to turn off the current. "You're doing well today. Let's hope we make it home."

During the summer, I spent many hours building a pyramid of rounds. By late August, I was ready to start splitting—an especially pleasant job thanks to my gas-powered log splitter, which I've named "Maxwell" in honor of Hu Maxwell, a historian and forester from West Virginia who wrote his most ambitious book, *A Tree History of the United States*, in 1923.

I like spending time with the mechanized Maxwell because he's an objective historian. He reveals past events with authority and precision and does so without expressing his opinion. Aside from occasional care and feeding, he makes few demands and never argues. I'd like to think that I'm similar to him in this respect. When I mentioned that to Priscilla, she said, "Okay, he can join me for cocktails, and you can stay in the field."

One warm evening, I met Maxwell in his usual summertime quarters: a flat corner of a small pasture close to the forest tree line. A few feathery clouds drifted overhead, their edges turned pink and purple by the summer's setting sun. Next to Maxwell sat a hillock of wood ready for splitting.

I pulled the cord to start his motor, and our work together commenced in a congenial choreography. I removed a hefty log from the pyramid, pivoted slightly with a graceful *rond de jambe*, and set the log on Maxwell's steel runner. His hydraulic piston, thrusting a wedge slowly into the wood, split it with strength and grace. I caught one piece with my left hand and tossed it onto the mound of split logs piling up on a nearby flatbed trailer. My knee wedged another piece against the runner until I lobbed it toward the flatbed in an elegant arc. I finished the sequence with a simple turn back to the pyramid.

Throughout our dance, my right hand rested lightly on Maxwell's horizontal, rubber-covered lever. A gentle pull ushered his piston forward toward the end plate; a slight push and the piston retracted into its hydraulic pipe. And then again, and again, and again. The steady puttering of Maxwell's small motor provided the musical accompaniment for our *pas de deux.*

In many instances, the bark showed few hints of what lay within; as the logs split apart, the tree's history was revealed—a history written in the long interior ligaments of the heartwood, sapwood, and cambium. Occasionally, my eye caught the markings of a story unusual enough to stop the dance and look more closely. On these occasions, the ligaments were disordered, their sinews twisted violently by the root of a branch whose exterior presence had disappeared long ago. The wood's interior grain was akin to water flowing around a river's rock—but frozen at a particular moment in the tree's life. Looking down at the split log, I saw an undulating river of wood, its contortions caught in the afternoon light.

If I'd sliced the trunk horizontally, I'd see the familiar form of a knot's face. But the vertical view told a more compelling story about the imperfections that had shaped the inner life of the tree—hidden imperfections exposed only in death.

Sometimes, the log's inner contortions created rotational shapes like a satellite's picture of a swirling hurricane. In one split log from a white oak, I saw a cyclone caught in a moment of intense ferocity, the wood's ligaments warped tightly in a distorted, silent spiral. No meteorologist had paid attention to that storm's story. Just me, a lone witness to a single perturbation in a long life. I held part of an oak tree that had seen many seasons—hard winters, dry summers, warm autumns, lovely springs. Its age was close to mine. Did my body, in its worn-out cells and dilapidated double helixes, contain the same testimonies to life's fray?

One day, Maxwell's piston split a log to reveal a pale-white wood borer cloistered in a blackened tunnel stretching from inside the bark to the tree's pith. The tunnel was littered with cake-like crumbs of chewed wood, residue of the larval stage of a wood-boring beetle. The tip of the worm—its head or tail, I don't know which—curled up suddenly. The evening light had disturbed its rest. I'd stolen its quiet solitude and, worse, its protection.

The worm itself was ugly: a groveling, gelatinous, floppy creature, a powerless cousin to Jabba the Hutt. As I stood gazing down, it twisted slowly in its tunnel. I'd ruined a bug's life and disturbed the natural world yet again. I was one more human who didn't foresee the implications of his actions.

I slid the wood borer from the log to my hand, felt its sliminess, and walked to a nearby stump. I set it down, hoping that some bird would find great value in an easy meal. A measly effort to make amends for its inadvertent assassination.

Maxwell was silent now because he'd run out of gas. Nearby crows cawed loudly. Had they spotted the bug or were they gossiping about a ridiculous human? I fed Maxwell his fuel, started his motor again, and picked up the next log.

By late September, I'd reached the bottom of the pile. The last few rounds were from an old pear tree we'd felled a year ago. Pear wood is beautiful and fine-grained, which makes it smooth to the touch. Splitting the remaining pear tree rounds took longer than usual because I kept stopping to admire the astonishing array of colors inside the wood. One quartered log had lines of deep red, sienna, and tan. Another was streaked with multiple browns: chocolate, mocha, and sand. Yet another had the lighter end of the spectrum: peach, rose, and salmon. The logs were akin to canvases painted with variations of a desert sunset.

When I finally split the last log from the pear tree, evening was settling in. A few gray clouds drifted across the darkening sky. I silenced Maxwell's motor and nodded goodbye. He'd been a reliable partner, a simple machine with a single purpose. His puttering faded away, replaced by the din of crickets. I picked up the final quartered piece of wood.

In the fading light, I could barely see its canvas—a deep red background with undulating lines of faint rose and subtle plum: a sunset caught in wood. When I got home, I showed it to Priscilla, who is by nature a talented colorist.

"Yes, that's beautiful," she said. "Don't you remember that I often used those colors when I was painting my silks? Everyone loved the scarves with the ochre tones. I got lots of orders for them, especially in the fall."

I kept that pear log on a bookshelf, but its colors faded after a few weeks. I threw it in the fire and imagined its smoke swirling though the chimney, drifting into the chilly night air.

Unexpected Outcomes

Henry

Shortly before my formal departure from professional life, I decided that after retiring, I'd write stories and essays purely for fun, rather than for clients and colleagues. How naïve was that? Writing is never easy—at least not for me. In retrospect, the notion of writing for fun was a delusion.

In the initial weeks of the wonderful and unsettling freedom that followed my retirement, I postponed sitting down to write. No reason to begin having fun right away when the future seemed endless. Instead, I caught up on my reading, started and even finished some long-postponed chores around the farm, and stayed up late to watch the stars.

But epiphanies often arrive without formal announcement.

One summer day a few months after I'd left my job, I was reading our local county newspaper, the *Hampshire Review*—a large-format, weekly rag. I have always appreciated local papers, especially because so many have disappeared in the face of social media, blogs, online sources of news, and many other forces in the publishing industry. At the time, the *Review* was a solid newspaper that typically won state-wide contests for its journalistic quality—thanks to a veteran managing editor from the Chicago newspaper world, a highly committed staff, a talented young reporter in her late twenties, and a small squad of volunteer contributors with varied points of view. (It's still going strong with different leadership and a few new staff members.)

As I sat on our deck skimming an article about a local family with several generations of farmers, the sound of our neighbor's tractor grew increasingly loud. He was mowing our fields directly behind our house to make hay for his cattle. He worked quickly because, I assumed, the forecast included the possibility of rain. What is actually involved in making hay? I wondered. *What happens if the rain spoils it all? What will he do then?*

I looked out the window at the tractor growling along the hillside and then back at the newspaper. Why not write something about haymaking or, even better, about h*ay makers*? Even good local newspapers need halfway decent copy, especially if it's focused on people in the community. If I could write about what I found interesting, maybe it'd be good enough for the paper. I was hooked.

My idea led me to record formal interviews with four commercial hay farmers working in our county. Based on the interview transcripts, I wrote a series of six articles for the paper. A professional photographer (and a good friend) agreed to take relevant pictures. The articles and the accompanying photographs were well received, even by veteran hay farmers who rarely read anything and who believe there's nothing new to say about something their families have been doing for generations. "Good stories," one of them said to me. "I actually learned something."

The hay farmer project taught me several lessons. One

of them involved the nature of community in a rural area like Hampshire County. The hay farmers had been willing to talk to me because they were proud of their work, their machines, and their knowledge. They hoped I'd tell their stories forthrightly, and I did because they had opened their hearts to tell honest tales. But along with their resolute pragmatism about farming ("just get the job *done*"), they made only passing references to discomforting facts about their futures: hayfields that would eventually become unavailable because of new townhouses, unaffordable costs of new machines, technological barriers to repairing old ones, changing customer demands that require new marketing strategies, and the cumulative effects of a vanishing labor supply on the farmers' aging bodies.

My interviews with the hay farmers also highlighted tension between their natural hospitality and suspicious insularity. Like many people whose family stories go way back in a community's history, the hay makers were careful about talking to someone they considered a stranger. I hadn't graduated from the local high school, hadn't logged a lot of time at the county fair, hadn't had a reason to attend school board or farm bureau meetings. To them, I was a likable fellow, a curious outsider willing to learn about them and their craft. But that was it. They'd never invite me over for dinner because they'd never consider me part of their community.

When the hay farmer project ended, the empty page problem surfaced like Melville's whale: the pure whiteness of my computer's blank document page struck panic in my soul.

To avoid that agony, I turned to walking alone through the farm and riding horses with Priscilla on the mountain behind us. By this time, a year had passed since I'd had an easily available community of colleagues and work acquaintances. Aside from Priscilla, two friends up the road, and a few more who lived in distant cities, I didn't have anyone to talk to.

For the first time since my youth, talking to imaginary friends became an inviting exercise. Priscilla had been schmoozing with her livestock for years. Why not start a quiet conversation with trees, rocks, and birds? Rather than simply talking to myself, I could gin up some dialogues with nature.

Out of this idea came a year's worth of monthly columns for the *Hampshire Review* in the form of "letters" to the natural world—imagined conversations with rain, wild animals, the wind, and other parts of the forest world around the farm. In a literal and overly simplistic way, these letters began a series of discussions with nature's representatives about their lives and values. I sought to talk directly to a few animals, our pond, and certain trees and discover what they might say in return. At its heart, this exercise forced me to confront my notions of nature and wilderness. Exactly to whom or what was I addressing my letters? And what was my place in the natural world, anyway? As a farmer (actually, an assistant to Priscilla, the true farmer in the family), my role was clear: tending to the livestock and pastures as respectfully as possible. In the woods beyond the pasture's fence, however, my role was far more uncertain.

Writing my letters led me to examine more carefully my longing for an alternative to the asphalt-filled, concretized, blaring city life. Was I destined to remain a city boy, always apart from the natural world? Or, by spending time in nature (whatever that means), could I become a part of it? The simple, if unsatisfying, answer was "yes" to both questions.

The more complicated answer is "it depends." To the

extent that I personified nature in my letters, it comes across as indifferent to the human condition and skeptical of my limited understanding of its complexity. Depending on where I look, it contains the beautiful and the brutal; depending on what I hear, it contains the song of the thrush in courtship or the whimper of the mouse in the owl's talons; depending on where I walk, it contains the sweet perfume of wildflowers or the stench of a deer's rotting corpse.

I settled finally on the simple acts of looking, listening, smelling, tasting, and sensing. I came to understand the value of experiencing the natural world first with my senses rather than my mind. Let me feel before I analyze and judge. Or, even better, let me just feel. Priscilla started encouraging me to do this soon after we married. I've gotten better, but I still depend on her to remind me that feelings, fully recognized and embraced, are a necessary part of being human.

For someone who had long depended on my mind alone to guide me, the effort to embody my experience was no small matter. Eventually, I was able to stand in the forest for a while and feel without too much thinking—a practice that many others have mastered in their pursuit of a spiritual connection with the life around us. In any event, it seemed as if I was beginning to learn the speech of the natural world, one that is infused with movement, sound, and scent rather than structured with comma, clause, and case. Robin Wall Kimmerer understood this matter years ago. In *Braiding Sweetgrass*, she writes, "Listening in wild places, we are audience to conversations in a language not our own."

After my monthly columns had run their course, the paper's managing editor emailed about a person who wanted to track me down: John D. Gavitt. "He's a pretty wonderful guy," the editor reassured me, and attached a short essay that the fellow had written about his own property, which he'd sold two years earlier. The essay, titled "Sacred Ground," was in the form of a letter written to the future owners of his property.

I began reading his essay-letter with great reluctance. Out of simple courtesy, I'd have to respond to the fellow and, inevitably, he'd ask what I thought about his piece. What if it was a terrible bit of writing: sappy dribble, embarrassingly personal, or wholly unintelligible? I'd have to find a way to be diplomatic.

But it wasn't any of those things. Instead, it was an eloquent testimony to the spiritual power of place in one person's life. In his letter to those who now own his land, he wrote, "The cycle of life and death in the natural world of this property continues year after year, as it will when you and I are gone. . . . This 'sense of place:' I suppose that's what it's all about for me and for others who believe so strongly in a particular chunk of this earth. It becomes so much a part of us that we will do everything possible to ensure that it will not be harmed when we're not around to care for it."

I called him the next day, and we found time for lunch a week later. We've become important to each other. A new friendship, wholly unanticipated, born through words.

Three years after retiring, one more surprise derived from essays written for the local paper: an invitation to join the board of directors of a local organization dedicated to environmental preservation, The Cacapon and Lost Rivers Land Trust. Like all land trusts, it aims to protect forests, farms, and rivers within a specific watershed by collaborating with landowners and farmers to establish conservation easements.

The invitation was startling because I have no technical knowledge or experience in any discipline relevant to the Trust's efforts (biology, environmental law, river management, forestry, or the like), and I'm no conduit to sources of financial support. But the Trust's mission includes outreach to the public as well as to landowners who might consider developing a conservation easement for their property. The staff needed help with newsletters, brochures, and other publications. I was happy to accept the invitation, suspecting that I'd learn far more than I'd ever contribute.

All the unexpected outcomes of my writing—the new communities and friendships it ushered me into—stem from the reality of our farm and the natural world around it. This place has both physical presence and symbolic power; for me, writing about the first invokes the second. I'll never fully grasp the land's many contradictions, surprises, and idioms. At best, my words capture my love for this place and tell of my conversations with its inhabitants. Other folk are welcome to join us.

Lucy

Priscilla

Lucy was a big and healthy doeling with perfect conformation, except for one thing: our dear gal had been born with extreme strabismus.

Her eyes turned outward in radically different directions. It was hard to imagine what she saw out of those peculiar eyes. Was it like constantly looking at two screens at the same time? To make matters worse, her head was turned to the left at a twenty-degree angle—not because of any bone malformation but probably because of the strabismus.

Because of her head tilt, she looked as if she couldn't walk in a straight line, and the quirky eyes didn't help. But she moved out to the pasture with her family and played with the other kids. Within a couple of months, the herd had accepted her as part of the new crop.

Lucy was a wonderful goat except for the strabismus, which put her in a special class. If she'd been puny or a hard keeper or if she'd had a bad leg, I would have seriously thought about putting her down. But she didn't have any other weakness that I could see.

Lucy and her twin sister, Mauzy, were smart and friendly in a rambunctious goat sort of way. They were on the big side for doelings and two of my most parasite-resistant kids. They had some of the best bloodlines in my Spanish herd. Their mother, Edith, was a fine, productive Spanish doe. Their sire, Elvis, was my favorite Spanish buck. And Mauzy was completely normal.

Where did the strabismus come from? The sisters' bloodline goes back three generations on my farm. I had never seen strabismus before in my herd. A few other reports describe this heritable condition in goats, but it is extremely rare. It's found mainly in cattle. The strabismus wasn't contagious, so I didn't have that worry. Its sudden appearance was a sad mystery.

Every year during kidding season, I kept track of the kids I'd probably keep in the herd and those I'd probably sell in the fall to other breeders to keep my herd's bloodlines balanced. But there were always a few kids whom I'd have to butcher because they were too small or too prone to disease or not impressive enough to join my Spanish goat conservation program.

Lucy presented a special problem. She was healthy with good confirmation and came from good stock. But did I want to risk her birthing kids with strabismus? And who would ever buy her? It would be unethical to sell her. Should I put her down?

If I did put her down, it would make sense to wait until she was at least big enough to slaughter for meat. But, in any case, I didn't need to make the decision while she was so young. The twins stayed with their mother and the herd through the following year.

In the fall when they were about eighteen months old, Lucy, Mauzy, and the other girls in their cohort were ready to be bred for the first time. Either I could send her to the butcher or put her in the field with my Spanish buck and other young doelings. After going back and forth in my mind, I decided to see how she would do as a mother. I figured waiting wasn't going to matter.

It turned out that Lucy had a normal, healthy pregnancy and an easy birth. She delivered a very nice buckling. He was completely normal, and she was a great mom. Mauzy birthed a healthy doeling and was also a great mom. Many times, Mauzy and Edith left Lucy to watch over all their babies while they went with the herd to graze—a strong vote of confidence in Lucy's skill as a mother.

But a question roared in my head: *Do I have room for another doe that will always be a question mark? Why would I treat her any different than any other goat with a limiting health condition?*

No matter how many times I preached her strength to myself and others, I couldn't imagine someone picking Lucy over all the does. Who would buy her when they knew the strabismus was heritable?

I don't know why, but I kicked the decision down the road again.

Later in the summer of Lucy's third year, a fellow called me about buying some Spanish does that were proven breeders. He and his brother ran a big farm in the southern part of the state. They had made their money in the cattle business and in oil and gas drilling.

In the search for another income avenue, the brothers were planning to become Boer goat breeders. The Boer goat market was booming at that time, and consumer demand for goat meat was growing rapidly. Boer was the meat goat of choice, and they wanted to make a name for themselves in that business—which meant they had a lot of money tied up in Boer goats. But they had learned that having a single breed was a risky proposition. They needed more genetic diversity in their herd.

One of the brothers had read about me and my goats in a farm journal. The article described how all my goats had names, were relatively friendly, grazed in pastures with guardian dogs twenty-four seven, and needed very little feed or help, especially as mothers. The article explained my practice of pasture rotation and my expectation that goats must do their jobs (like birthing kids) in the pastures unassisted.

During the call, the fellow wondered if I would talk with him and his brother about my goat management practices. They were nice guys. They asked the right questions and said they would pay for my time. I was thrilled at the possibility of spouting off my views on natural animal husbandry. So, we set up a time to meet in person on my farm so they could see my operation.

Boer goats were known to be sorry moms and hard keepers. I expected the brothers had a bunch of expensive Boer nannies who had highfalutin papers but couldn't birth their own kids or even get bred without human help.

At the meeting I learned I was right. The brothers' problem was a herd of expensive Boer nannies from top-line registries who couldn't perform their only job.

With the financial investment the brothers had made, they had to work with the stock they had. They investigated developing a surrogate program for their goats. A surrogate program in livestock is much the same as with people—a doe, or host, is implanted with a fertilized egg and, ideally, carries the baby to term.

The brothers decided they had the resources to start such a program on their farm. All they needed were dependable does that could carry Boer babies to term and mother them until weaning. They wanted some of my Spanish does to be hosts. At the meeting, I also learned they had decided to get into this game because of their children. One brother had three children; the other had two. The brothers had grown up working on their parents' farm and showing cattle. They hoped to get their children interested in showing their home-grown Boer kids. But, to get to that point, they needed goats that were gentler than Boers and that could ease the children into handling animals. It would be nice if those goats could also be in the surrogate program.

During our conversation, one of the brothers mentioned that his youngest daughter, Bonnie, then ten, had autism. Bonnie's dad spoke of his happiness as he watched her work on the farm and lose herself completely in her jobs, especially the chores that included the animals. In fact, the animals were one of the few things that held Bonnie's attention and gave her joy. He hoped that showing goats might be a good activity for her.

The family knew of Temple Grandin, a leading animal behaviorist who is famous for her work in promoting and designing more humane ways to operate a slaughterhouse. Having autism herself, she also is an outspoken proponent of autism rights. In Bonnie's household, Temple Grandin is a hero.

Somewhere in the conversation, I realized I wasn't sure about selling my Spanish does to these brothers. I questioned if this was the right decision for a heritage breed. But I had an interest in what they were trying to do with their children, and I saw a possible opportunity for Lucy. Ever since she was a kid, Lucy had enjoyed being handled; she loved affection from everyone. Lucy could be a good fit for Bonnie.

Goats, like all animals and all people, need a purpose. Sometimes it involves breeding; sometimes it involves providing milk or their body as meat. Lucy's purpose would be twofold: surrogate and partner. She'd never pass the strabismus to her offspring and, just as important, she could help a little girl find her passion.

I told them I would think about their story and get back to them the following week.

That night I talked over the pros and cons of my plan with Henry, who listened with good questions that made me think out my proposal to the brothers.

After a while, Henry gave me a hug and went to bed. I sat in the darkness, pondering my decisions. How many does should I sell them for surrogacy? Would Lucy have a good future?

The next week the brothers and I worked out a deal. The first group would be eight proven does: some Spanish-Savanna crosses and some full Spanish. They would not have any registration papers, but they would have up-to-date health records, and I would pay to get them vetted.

As I did the paperwork the night before the first group left, I came across Lucy's birth records. I looked again at the fact that I had three generations of her bloodline on my farm. The question of the origin of the strabismus was still a puzzle. It never appeared on the farm again. I folded the paperwork and bill of sale and stored it away with the rest of her family's records. All three generations had been in my conservation breeding program. Lucy belonged to that bloodline with or without strabismus.

On pick-up day, the brothers brought their children to see my operation and help load the new goats. The children wanted to see the big herd and asked me the names of various goats. We rode to the back pasture and walked around.

The children enjoyed the visit, and the guardian dogs were thrilled having the children in the pasture. All the dogs were friendly in their quiet, giant way, wagging their tails and following the children as if they were newborn kids.

I saw Bonnie take an interest in Lucy as soon as they met. Bonnie put a halter on Lucy and led her around. She was never bothered by Lucy's appearance or unusual way of walking. Maybe Bonnie and Lucy both knew the match was going to be good—as did Bonnie's father.

The children led their goats to the trailer, talking to them in kind and gentle voices as they went. I smiled as I listened to the children giggle and make their plans. They carefully loaded the new goats into the trailer. I felt a strange mixture of happiness at Bonnie's joy and sadness at Lucy's departure.

The next week, the brothers called me with news that my goats were a hit. Apparently, the children had talked about their new goats the whole way home and decided which goat belonged to whom. To the brothers' surprise, all the children helped the goats settle into their new home. From the first night onward, the Spanish goats had names, water buckets, and fresh hay in their mangers.

The brothers came back once a year for a few years, even if it was just for a couple of does. Over lunch in the pasture, the brothers and I had good laughs at stories of the dogs, kids (two legged and four), and the wonders of farming and parenthood.

For my part, I loved getting caught up on the stories of the Spanish goats and their gang of youngsters. All the goats lived out on fields with dogs twenty-four seven. Tutored by my Spanish goats, the Boer nannies had learned to be better mothers and were doing well. Some were even birthing in the pasture unassisted. After a while, everyone in the goat herd had names. The brothers knew that tickled me!

I had told Bonnie's dad that Lucy was kind to everyone. He said the strabismus didn't matter to Bonnie. She and Lucy understood each other. With commitment and hard work, Lucy and Bonnie went on to show in the local livestock festival. Bonnie was obviously proud of Lucy, and Lucy seemed to relish the attention and stayed calm despite the fanfare of the festival.

Eventually, the brothers put Lucy in the surrogate program. The plan was for Bonnie to help bring Lucy's kids along—a plan that seemed great to Bonnie. Soon enough, Bonnie had a beautiful Boer doeling from Lucy to work with and show. They even won second place in their first show. Back at the farm, Lucy always followed Bonnie around like her big dog. They were their own tiny herd.

The farm chores and animals had become Bonnie's passion. Her dad was convinced she'd have a good future on the farm. Unfortunately, he did not live to see much of that future. He died of a cancer he'd been battling for close to five years. Before his death, he and Bonnie worked every day on farm chores. He had told me often about the joy he found watching Bonnie pull Lucy around those first few days and couldn't believe that

Lucy had so much patience. He delighted in how his daughter became wonderfully lost in her time with an animal she loved and that loved her back. Perhaps Lucy felt that Bonnie was her plan, her purpose. It was very clear that Bonnie was convinced Lucy was her priority.

Bonnie and her dad worked together for about four years before he died. After his death, Bonnie and her mother decided to stay on the farm and help keep the goat business going. The last I heard, Bonnie and Lucy were still a team.

Izzy's Bridge

Henry

"It's still here!" Izzy yelled. "My bridge is still here!" Her shout resounded through the trees.

I was walking in the woods with my twin seven-year-old grandchildren, Izzy and Charlie. The April morning was exquisite: a bit cool, a slight breeze, and sun pouring through sparse leaves onto ground still wet from an overnight drizzle. Three weeks earlier, during a previous visit to the farm, Izzy had built a wobbly bridge with a half-dozen branches collected from the forest floor. It crossed a small run bed downhill from our wildflower meadow.

I hadn't yet checked the site, and before we started our morning trek, I'd told her about the thunderstorms of the previous week, mentioning that the water had come up quickly all around the farm. Izzy had been worried about her bridge and rushed to get ready. She brushed her shoulder-length brown hair quickly, ignoring her mother's suggestion to put it in a ponytail, and put on her new sneakers. "Aren't they pretty?" she asked before dashing out the door.

Meanwhile, Charlie sat on the mudroom floor pushing his feet into rubber boots that were slightly too small and fending off his mother's attempts to help. When he stood up, he grabbed his favorite cap ("my dusty Husky hat"), adjusted it just so over his short brown hair, and leaped down the stairs after Izzy.

"Come on, Charlie!" Izzy hollered. "We've got to go see what's happened to the bridge."

Running to the spot where an old logging trail jutted away from our dirt road, they slowed, waiting for me to catch up.

"Keep going," I huffed. "I'm coming."

Charlie ran on, following the trail as it dipped down and then up to a small ridge. Izzy took my hand and we walked in silence.

"It's hard being a kid," she sighed. "You have to do all these things parents and teachers want you to do. Just so they're happy."

I had no idea what had prompted her comments, so I didn't say anything.

"I wish I knew what you know," she continued. "I wish I was an adult. Then I'd know things."

I could have asked, "What do you want to know?" But that sounded too adult-like. Instead, I replied, "I like knowing things because it helps my imagination. The more I know, the more I can imagine."

"There you go again, Papaw. Saying things I don't get."

"Well, for example, take that tall oak tree over there. I know it has really deep roots, so I can imagine what might happen when a big wind comes along. The tree will sway and maybe its top will break but it probably won't fall. I can't see the roots, but I can imagine how they're like fingers holding the tree in the ground."

She let go of my hand, and I walked on until I realized she hadn't kept up. I looked around and there she was: standing stiff, her arms straight down, eyes closed.

"Izzy, what are you doing?"

"I'm imagining I'm an oak tree and my feet are way down in the ground. My toes are holding onto the dirt. I can't move."

At that moment, Charlie yelled, "I can see the bridge!"

In a flash, Izzy was running full tilt down the trail. I followed at a slower clip.

"It's still here!" she screamed as I caught up to them. A bundle of long sticks and dead limbs lay across a shallow run bed. The branches—some more twisted than others, some thicker than others, some more rotted than others—stretched from one side of the run to the other. They cleared the water by a few inches.

Izzy jumped over the stream to examine the bridge from a different angle. "Look!" she shrieked. "There's a hole under this side. It's like a cave! That wasn't there before."

She pushed the branches closer together and stepped back.

"Can I walk across it?" Charlie asked.

"No!" She scowled at her brother. "Of course not. I've got to make it stronger."

Charlie shrugged and wandered downstream to explore several large rocks. Izzy walked upstream in search of more sticks. I leaned against a stout oak, gazed through the sun-dappled, early spring leaves, and savored the land's sculptural beauty. Tethered by my senses, I embraced the world around me—the multitude of greens, the hillside's sensuous curve, the wind's slight rustling, the scent of damp earth. Reality, fully felt, was enough. I was delighted by the pleasure of letting my grandchildren enter whatever enchanted world they wanted to invent.

Given freedom from adult guidance, children will play with whatever is at hand. I've watched children from impoverished parts of my hometown, New York City, build small sculptures from threads of shredded tire, shards of glass, and broken bricks. That's what was available and so that's what they used. They had gone about their task with the intensity common to seven-year-olds, but with the wariness of kids who'd spent a lot of time on city streets. Hunkered over the hard sidewalk, their bodies were taut and their eyes watchful. They were vigilant for danger, worried another kid would come along and bust up their work, make fun of them, or push them away and take their carefully collected materials. Past experiences made them hurry; sometimes they'd kick apart their small creations before anyone else could.

Izzy and Charlie live on a quiet street in the Maryland suburbs outside of Washington, DC, and don't have to worry about the dangers of city life. Accompanied by one of their parents, they ride bicycles through their neighborhood, have playdates, go to birthday parties, join soccer teams, and complain about limits on screen time. Their lives are safe, well-protected, and highly scheduled. And like most children of cities and suburbs alike, they rarely have an opportunity to wander around the natural world, taking their time to explore what they discover in the absence of hovering adults.

"Look," Charlie hollered. His boots had filled with water and his pants were wet to his knees. "Look! The stream's going faster." He had used a stout stick to lever several heavy rocks downhill, widening the stream's channel. The water gurgled loudly over the new formation of stones. He bent down, reaching under the water for something, intent on the next phase of his project.

"Charlie! Your pants are soaking!" Izzy yelled as she lugged several large sticks toward her bridge. "Mom's not going to be happy."

Charlie ignored her. He was busy liberating the stream. She threw the sticks on the ground and then carefully placed each on the existing bridge, making it wider but not much stouter. She crouched to assure herself the cave-hole was still there and went off to search for more fallen limbs.

When they were eleven and thirteen, my two sons built a treehouse close to where I stood. I offered to help, but they insisted on building it themselves with old boards stored in the barn. They lashed the crossbeams to the trees with rope and used an exceptionally large number of nails to secure the floorboards to the beams. Its rope-and-wood ladder hung over the edge, taut as a spider's line in the wind.

The boys used it for two summers—as battleship, fortress, airplane, and secret hideout—and then abandoned it for other pursuits, like building a paintball arena for their buddies. The treehouse lost a few floorboards one winter and a side rail the next.

On a visit during his late twenties, the older of the two (and the treehouse's primary architect) noticed that a single board dangled from the tree's trunk by a few remaining strands of rope. He was surprised that anything was left.

"I can't believe you let us build the treehouse so high," he said. "It's twenty feet off the ground! We had to jump from one tree to the other while we were building it. And you let us do that?" The fearless imagination of a thirteen-year-old had transformed into the cautionary risk-assessment of an adult. It happens to most of us. I was glad—and sorry—he'd grown up.

Charlie, with his mission completed, sloshed upstream to the bridge. "Can I walk on it?" he asked. "What's a bridge for if you can't walk on it?"

Izzy placed her final branch across the stream. "It's my bridge, and I'm going to walk over it first. Follow me."

Izzy was halfway over the lopsided two-foot-wide bridge when the branches shifted and settled. She stopped, balancing herself as if teetering on a high wire and screaming with fear and delight. "Charlie! Help me. It's a bridge quake."

Charlie stepped on the bridge, which settled some more, prompting screeches all around. He took a few more steps, and then they both ran, stumbling across the wobbling sticks. Izzy reached the other side, turned to catch Charlie coming across, and laughed as they tumbled to the ground.

"We made it!" Izzy hollered. They stood up and hugged each other. "We gotta check the cave!"

That day, in our woods, the twins were in a land that was both a living sculpture and a repository of objects to be found—simultaneously art and materials for making it. They had taken what was there—water, rocks, and wood—and changed them effortlessly into sluice, cave, and crossing. They had no need of adults to find enchantment in the land.

I finally had to tell Izzy and Charlie we had to go. It was way past lunchtime, and their mother would be worried. We headed to the old logging road that would lead us back home.

"Do you think the cave will still be there when we get back?" Izzy asked, this time in a somewhat subdued tone. "It's a nice place for something, like a mouse family."

"Maybe a snake family," Charlie volunteered.

"No, Charlie. Snakes eat mice. What do you think, Papaw?"

Charlie hustled ahead of us. "Our birthday is next month."

"Stop interrupting, Char. But he's right, Papaw. We'll be

eight when we get back here. So, what do you think? Will the cave be okay?"

"Well, Izzy, I don't know. We'll just have to come back and take a look."

"I hope it'll still be there," she said, her voice a bit wistful.

Home Before Breakfast

Henry

The telephone rang about eight o'clock on a late October evening. Our neighbors, who rarely speak to us, were calling.

"Your horses, they're in our front yard," the wife said with smug irritation. "They're under the lights, eating our grass. I'm looking at them right now, out the window here."

"Really?" I replied. "That's odd. They must have gotten out through the back gate."

"Figured so. You comin' to get 'em?"

"Of course. Thanks for the call," I said.

"Figured you'd want to know."

"Right. We'll get the lead ropes and be right over."

"You know, they might not be here." I heard her chair creak in the background. She must have shifted in her seat for a different view. "Fact is, they're moving around. You wouldn't want 'em to make the blacktop. A driver's liable to hit 'em. They drive fast these days."

"Like I said, we'll be right over."

"They looks like there's no halter or nothing."

"Yeah, I know. We'll bring their halters too."

"We wanted you to know."

"Right. Thanks for calling." I hung up before she could say anything else.

Priscilla and I climbed into our small utility vehicle (known in the family as the "ute"), drove to the barn, and grabbed the horses' halters and ropes. We headed to the back gate and, sure enough, saw in the dim light of the ute's headlamps that the gate was ajar. We left it open in case the horses decided to retrace their steps.

We took the cut-through into our neighbor's yard and motored from side to side. No horses.

We aimed the ute's headlamps down our neighbor's driveway and shined our flashlight into the surrounding woods—futile efforts to pierce the night's gloom. We might as well have been using a match to find a crafty thief hiding in a dark cathedral. Whistling and calling out were fruitless.

"Wid, come on back; Gabe, we've got some treats. Wid! Gabe!"

We took turns getting out of the ute and walking a few yards into the woods, hollering for them over and over. No horses.

I was quite certain they were standing in the forest beyond the reach of our headlight, listening to us, watching the flickering flashlight. Wid, the smarter of the two, probably pushed open the gate and decided to have an adventure. Gabe, far more timid but loyal to his friend, would follow.

Priscilla's been around horses all her life. When she was four years old, she rode regularly, and at age seven, her father stuck her on horses to train them. She was only thrown a few times because she learned their language. She has that talent even now: horses accept her as one of them. Within her horse-riding community, she's celebrated as exceptionally knowledgeable and intuitive. And like Jonathan Swift, she generally prefers the company of horses to people. She reminds me of that fact whenever our arguments heat up.

That night, she was worried the horses would get to the blacktop and run up the hollow road into someone's yard. We might never see them again.

We motored down the neighbor's dirt driveway to the road and drove a half mile in each direction. No horses. We stopped a few times, turned off the ute's motor, and listened for the horses' clip-clopping on the blacktop. Nothing but the scratch of a few far-off crickets. One car approached from behind us, flicking its high beams. It tooted and passed, picking up speed as it disappeared into the darkness.

"If they don't come down to the road, they'll be okay," Priscilla said as I steered the ute back to our house. She was still calling out every so often.

I'd learned to ride in my early fifties, motivated mostly by Priscilla's claim that we'd need something to do together after the children had left home.

"I'm not giving up horses," she announced in her usual diplomatic way. "Learn how to ride or we won't be seeing each other much."

I figured I'd become a decent rider in two years if I spent a lot of time in the saddle, took a few lessons, and learned some horse talk. I didn't anticipate the pain and agony: leg muscles sore from wrapping around Gabe's broad barrel, shoulders tightened too often from the fear of falling, ankles aching from bending at awkward angles. The worst part was the embarrassment. Priscilla is a gifted horsewoman, but as a person trying to teach her middle-aged spouse a sport that, to her, was as natural as walking, she came up a bit short. She couldn't fathom my inexperience. When I needed a calm voice suggesting I hold my elbows closer in, she'd be chortling at my flopping arms. But after two years, as I had guessed, I did become a decent rider. I could last for six hours in the saddle and even enjoy it.

It was close to midnight when we gave up searching. After returning to the house, Pris went to sleep but I lay awake thinking about our horses—especially Gabe. It was hard to imagine life without him. I hadn't realized he'd come to mean that much to me.

What exactly was so important about my relationship with this horse? I was surprised that I'd never asked myself that question. Joni Mitchell might have been right after all: "You don't know what you've got 'til it's gone."

Gabe and I had spent a lot of time together on mountain trails—a few hundred rides during the last five years—walking, gaiting, cantering, and full-out galloping over all sorts of terrain. For him, I suspect our relationship was an uncomplicated affair. I was his human, a skinny two-leg who gave him food and treats and sometimes bad-tasting medicine and who often climbed on his back. He knew me and he trusted me. That was that.

For me, he wasn't just any horse. Riding him meant entering a liminal space between the wild and the cultivated. There I'd be, astride a thousand-pound animal capable of tossing me from his back, kicking me hard, and leaving me on the road. But he wasn't going to do that because we'd made an agreement—he'd take me anywhere as long as I kept him safe. It's almost the same promise that humans have made with the whole of the natural world—it'll give us life if we protect it. As a species, we don't seem to be keeping our promise.

Sometimes Gabe wasn't sure I could keep my side of the agreement. He'd stop at a wooden bridge he'd never seen before, convinced I was asking him to walk us toward death itself. If he caught a bear's scent, he wouldn't budge no matter how hard I kicked at his ribs. If a turkey suddenly took flight through the trees, he'd make a swift 180 degree turn that would almost land me on the ground. At those moments, his body would tense like a locked spring. He was ready to bolt and, if he made the decision to flee, there'd be no stopping him. Through my legs and butt, I'd feel the surging, primitive instincts of a wild and vulnerable animal, instincts that his species has never lost in all the years of human contact.

And when I let him gallop at full tilt, the wildness swept over us: the initial lunge, leaving the tame behind; the immense power from his hind legs, propelling us into the next moment, the next length, the next swirl of joy in his unrestrained, uncultivated, uninhibited abandon. I'd be holding on to the reins of the wild, the wind rushing by my ears, the boundaries between him and me blurring as I leaned forward, trusting that this horse, my horse, was loving as much as I was the camaraderie of flying together toward the feel of freedom itself.

None of these thoughts helped me fall asleep. I could not believe that I wouldn't ride Gabe again, that we wouldn't be partners again on mountain trails, that I wouldn't again feel his wild power under me. Finally, weary from the fruitless searching and the constant worry, I dozed off.

At first light, I climbed out of bed, dressed quickly, and left the house. I'd taken three steps from the front door when I saw the horses. They were right there, grazing on the deep green grass in our front yard. Gabe lifted his head and looked at me, as if to say, "Hey, good morning. What's wrong?"

Wid raised his head too. "Breakfast time?" he seemed to ask with amiable nonchalance.

As we all walked back to the barn, Gabe swung his big head back and forth slowly, asking me, "Drama? What drama?" He nudged my shoulder. "You'll forgive us, right?"

While they ate their morning grain, I fixed the gate. There'd be no more nighttime adventuring for a while, at least not through that gate.

I like to think they came back, and would always come back, because they have some affection for their humans. That might be so, but they also came back for breakfast and the safety of their pastures.

They had almost certainly spent the night grazing in the nearby forest. With the practicality of seasoned trail horses, they had taken advantage of an opportunity for freedom, a night to enjoy the shrubs and grasses that don't grow in their pastures. But they also would have remained alert to the forest's potential threats: bobcats, bears, and unfamiliar humans with ropes. I guessed they hadn't slept much. If they hadn't returned to us, the horses would have explored more of the world outside their pasture's fence, but, in their vulnerability, they would have paid the price of greater vigilance.

It's an old story, isn't it? True for horses and humans both. To be free is to be vulnerable. To be wild is to risk.

Quiet Times

The Pond

Henry

On a chilly, overcast October afternoon, an elderly friend and I climbed the steep hill behind the house and ambled toward our pond. The morning's cold rain—the epilogue of a violent storm that had pummeled the mountains all night—had finally drizzled to a stop.

"I'm glad we're taking this walk," my friend, Maurice, mumbled as he caught his breath at the hilltop. He wobbled a bit. I reached out to steady him, but he walked forward, dodging my arm with deliberate bravado. "You've talked so much about the pond, and you've never bothered to show it to me."

"Well, it's no Walden Pond, but it has its own small charms," I replied, wondering whether he could manage the fifteen-minute trek. "Maybe I should find us some walking sticks."

"Nonsense," Maurice replied. "I'm not that far gone. Get one for yourself if you need it."

We strolled through the pasture's high grass, our trousers and boots wetter with every step. I was glad my irascible friend had insisted on walking to the pond because it was one of my favorite spots. And yet, because of relentless farm chores,

I rarely took time for a visit. It was easy to ignore, and I often failed to remember its importance and its beauty.

Long before my family became the pond's custodians, a farmer created it for his cows; our goats, after drinking their fill, often sat in the shade above the pond. But, for me, it was far more than a simple watering hole. Lying behind a sloping berm, it's an unassuming stage for the natural world's daily routines and unexpected spectacles.

Like any good local bar, the pond gathers in a gang of nocturnal regulars: deer, bobcat, coyote, fox, fisher, bear. They may not appear at the same time but, sooner or later, they'll all saunter in for their usual drink. In the large scheme of things, they need each other as much as they need the water.

The pond is for the birds, too. Priscilla and I sometimes watch red-winged blackbirds building nests in the high grasses at its edge; twice a year we witness geese spiraling down for a stopover during their travels; and on occasional summer evenings, we appreciate how effortlessly the wing tips of the swooping, insect-chasing swallows carve evanescent curves on the pond's smooth surface.

A family of water snakes usually coils up under an overhanging thicket of multiflora rose. When the mother leaves her nest, her head periscopes above the surface, followed by a traveling trail of undulating ripples. Always, at this moment, the resident frogs fall silent.

Of all the pond's inhabitants and visitors, my favorites are the damselflies: ebony jewelwings, fragile forktails, seepage dancers, and others. What a cornucopia of shimmering colors! What a wild collection of names! A novice odonatologist's dream.

I zipped my jacket a little higher to keep out the chilling

breeze. No spots of color would be whizzing over the pond. The butterflies had disappeared weeks ago.

As we climbed the berm, my toes were small, wrinkled pickles in a sloppy jar of brine; my nose was icy from the cold; and sweat beads trickled down my ribs. Maurice—flushed, happy, and exhausted—rubbed his hands together to warm them. We finally arrived at the berm's crest and, to our utter surprise, found ourselves gazing down at disaster. A slew of dead fish—small bass, catfish, a few carp—rotated in a slow circle on the pond's surface. Blown by the wind, they floated around and around like a melancholy merry-go-round of death. We stood without speaking for many minutes, absorbing the eerie scene. I struggled to understand that virtually all the fish must have died. A small, innocent pond, just minding its own business, had suffered a profound catastrophe.

Maurice finally broke the silence. "Well, that sure ain't pretty," he said. "It'll start stinking bad if the weather gets hot."

After a few more minutes of pondering the grim scene, we watched Priscilla drive our old truck across the pasture. She'd figured we'd be wet enough to want a ride back. I gestured for her to come look. She frowned, put the truck in park, and came over.

She hadn't liked our neighbors for years and said immediately, "I bet a neighbor threw some poison in the pond. Come on, let's go home. We can't do anything about it now. And you guys need something to warm you up."

Back at the house, we removed our boots and socks and warmed our feet in front of the wood stove. Priscilla offered to make us an Irish coffee. It's one of her specialties; she always gets the layer of the cool, whipped cream just right, allowing it to rest gently on top of the hot black coffee and bourbon.

But Maurice didn't like sweet drinks. "How about a glass of dry white wine?"

With a bottle of appropriate wine at hand, we discussed the dead fish. Maurice suggested that an underground leak in the mountain had somehow infiltrated the water with methane gas. I volunteered to call the Department of Natural Resources and report the problem. Priscilla wouldn't budge from her position. "I'm telling you, someone poisoned our pond."

For once, Priscilla's theory turned out to be wrong. The next day I learned from a fellow at the local Department of Natural Resources that fish kills in small ponds are rare but not impossible in the late fall.

"It's been raining hard there, right?" he asked. "And it's turned cold, right? And, if your pond is like most ponds around here, it's got a lot of muck on the bottom, right?"

I replied that he indeed was correct on all three counts.

"Well, then, your pond probably had a temperature inversion."

He went on to explain in considerable detail that ponds typically have layers of water that vary in the amount of oxygen they contain. The lower layer contains less oxygen because it's too dark for photosynthesis (which produces oxygen) and because the bacteria that break down the accumulated organic muck at the pond's bottom produce mostly methane and sulfur (but not oxygen).

"So, fish typically stay in the pond's middle and top layers, where there is more oxygen compared to the lowest layer. Got it?"

"Yes, I get it," I said.

"So," he continued, "heavy rains and freezing winds can stir a pond's layers, right? And that's probably what happened to your pond: The upper layers suddenly lost their oxygen and the fish suffocated. It can happen overnight. That fast. It's called a temperature inversion because the top layer gets cooler and the bottom gets warmer and the oxygen gets thinner all around."

I smiled to myself. "It's like a devilish spirit took the wind and mixed the layers of a large Irish coffee, right?"

"I've never had an Irish coffee," he replied matter-of-factly, "but, yeah, I think that's about right."

A few days after the fish kill, I walked back to the pond. The day was dry and still, and the sky cloudless. The dead fish had vanished, either eaten by critters or submerged in the pond's mud. The surface was like glass again. As usual, the water was clear for a few inches before it faded into blackness; the pond's dark decaying muck remained out of sight.

"Water and meditation are wedded forever," Herman Melville wrote more than 170 years ago. He was thinking specifically of people who, standing at the edge of New York City's famous island, gazed toward the harbor and the rolling waves of the sea beyond it. But his observation applies anywhere. Most of us still let our minds wander as we gaze at oceans, rivers, and ponds, meditating on whatever troubles we face.

For me, the power of my pond lies in its inconstant surface and the obscurity of its depths. Its character is corrugated. Smooth one day, rough another. A visible face with an invisible body. Some days, clouds float on its quiet top. Some nights the pond's round, still surface is itself a single unblinking eye, right there, staring upward, reflecting the light of a thousand distant stars. And behind that face, that eye, lies a host of secrets. I'll never quite know what's going on below.

The pond embodies mystery. And I'm thankful for that.

I've become comfortable with the existence of mystery in my life. I feel no reason or compulsion to explain it. Yes, I'm glad to know how the fish died, but I'll never know quite why it happened. Death is a mystery itself. I'll never know what lies beyond the veil of life, and I'm not concerned about it. God, the universe, time itself—they're all mysteries. Yes, new insights into the world's workings are essential. I trust that humans will never stop discovering new things. But understanding the "what" and the "how" differs from wondering about the "why."

Sitting by the pond while I'm figuring out life's meaning offers a pleasant distraction from farm chores, but my philosophical qualifications are too thin to make much headway and Priscilla no longer puts up with my ramblings. "Why don't you go read *Moby Dick* again," she says. "How many times will it be? Fifteen?"

When her chores are too many or too demanding, Priscilla opts for having fun instead. And what better place than the pond? In addition to all its visuals that claim my eye and its mysteries that grab my mind, the pond invites play. Our grandchildren let fly their carefully gathered stones and the pond kindly responds with a pleasing "plop." Our dog leaps after a pitched stick, returning it with pride and hope for yet one more watery pursuit. On hot days, our sweaty horses step in, not only for a drink but also to roll in its cool waters, sometimes even with riders still in the saddle. When the pond's iced over, I can't resist walking out despite the risk of a slippery, antic dance. The pond always has another gift to give.

In the May that followed the October fish kill, our friend, Eddy, stopped by our house, carrying a large bucket. It held two fish splashing and thrashing against each other.

"We were fishing this mornin' at the river. Got some catfish," he said. "Thought you might want to put 'em in the pond."

Eddy and I drove to the way-back and slipped the pair of catfish into their new home. They swam off, their big whiskers and tails disappearing quickly in the muddy water.

Later that summer, Maurice visited the farm again and wanted to know what happened to the pond. I told him about the temperature inversion and the catfish.

"Let's go see the pond," he said. "But you'll have to drive me. I don't feel too good and I can't walk like I used to."

As we looked down from the berm, a few goldfish flashed on the pond's surface. At its far edge, a snake wiggled through the water, leaving a trail of ripples. We didn't see the catfish. If they were still there, they'd have been at the bottom.

"Funny how life just keeps coming back," he mused. "You think everything dies, but it doesn't." He gazed at the water for a while, perhaps meditating on his own mortality. He must have sensed—accurately, as it turned out—that his death was near.

"Come on," he said after a while. "Let's go look at your fruit trees. You need to prune them this fall, and I'll tell you how to do it."

"Maurice, I think I can handle it."

"You don't know what you're doing," he replied, reverting to his ornery, life-loving self. "Shut up and let's get to work."

On a Summer Breeze

Henry

One day in July, a summer breeze caught me working in the new meadow. It came around the tree line while I was leveling old stumps with my chainsaw, sweating in the mid-morning sun. When it brushed my face with its gentle touch, I smiled, as if I were a young man kissed by a secret lover. The slight breeze, scented with honeysuckle and fresh-cut grass, sashayed around me. I took off my shirt, sat on a stump, and savored its fragrant warmth.

As if I were on a stage with an audience I could not see, I said to the breeze, "How beautiful you are today." With a twirl of its large crinoline skirt, it swirled the air around me. A soft flirtatious hand swept along my bare back.

"I see you've come to seduce me."

The breeze disappeared, fleeing the stage in a huff, as if I had insulted it. Or maybe it was just playing hard to get. I sat in the hot sunlight hoping another member of the cast would show up to keep the performance going. It's a familiar conceit of mine. As I work on the farm, I'm often a solitary player on a rough and sloping stage, one actor among a sometimes clownish troupe of domesticated animals performing in a wayward drama drafted by an amateur playwright.

In my show, we all have a role, including the elements and animals of the natural world. Summer breezes show up often in my productions because I've been their passionate fan for a long time. Each one is grace in its most sensual form, love made palpable.

The overeducated, skeptical part of me is embarrassed by such romantic claptrap. A breeze is moving air, a simple draft, a part of nature that has nothing to do with gauzy human emotions. By asking inanimate matter to be more than it is, I'm guilty of mawkish sentimentality. A tempest is just a tempest, after all. No reason to think it has "meaning."

And yet, humans are meaning-making creatures. We see in nature's performances our own terrors, lusts, and yearnings. Grand views elicit our strongest feelings—how else to explain the traffic at Yellowstone—and many of our most compelling memories lead us back to important places in the natural world: a field of flowers, a hiking trail, a quiet beach, a park bench at night, a campfire on a cool fall evening.

My inner skeptic isn't convinced. "It's a ridiculous conceit," it tells me. "This notion of life as a grand play. It's just abstract mumbo-jumbo, hopeless hokum. You and I are no actors. At its best, this thing called a summer breeze provokes mind-numbing lyrics: 'Blowin' through the jasmine of my mind.' Really? That summer breeze you think you love is just a corny prop for the human imagination."

Maybe. But the breeze exists. Others feel it, not just me. It has physical substance, which is more than I can say about the characters in my own psyche. The summer breeze means something to me because it's shown up—it's walked on stage— at important times in my life. It appeared when school was over in the Junes of all those early school years, and so it means freedom. As a twenty-year-old standing in its fragrant breath, I watched a full moon shine on a far-away land, and so it means enchantment. I kissed my first girlfriend when it sauntered along a Massachusetts mountain top, and so it means wow.

And that's the point, isn't it? We're all in this play together: humans, animals, trees, rain, wind, stone, and the other elemental parts of the natural world. Those elements—the fierce storms, the dappled shade against the mossy rocks, the fragrant wildflowers—are more than just part of the set design. They're characters in the play; they have their own roles and dramatic flourishes. They have their speaking parts, too, if we listen.

In one of the most urgent dramas of all—the fate of our collective civilization—the climate screams its lines, "Listen to me. Change is here. Bigger changes are near. Who among you will save us?" The violent winds, the rising seas, the savage, scorching droughts—they've joined the troupe, they have a message, and they compel us to respond. If we play our parts poorly, some of us won't be on stage for the next act.

As I stood to resume my work, the leaves of a nearby oak rustled slightly. The breeze was back. But this time, as it swept past me, it reminded me that I was lucky. I wasn't standing on a sidewalk in the city of my youth, inhaling the fetid, urine-soaked air blowing up from the subway grate below my feet. I didn't have to smell the musty stench of decomposing drywall long after the flood waters were gone. I wasn't downwind from burning chemicals.

"Let's not get too dramatic," that skeptical side of me pleaded. "Yes, you're lucky. But you've been in bad places and now you're here. At least you're trying to appreciate what's around you. Anyway, there's a lot of randomness in life. Maybe there's no script at all. We're just making it up as we go along."

"Maybe," I replied and yanked the cord on my chainsaw. It roared to life and bit into the next stump. Not to be outdone, the breeze caught the sawdust and swirled it past my face, as if to say, "Don't forget me now. . . ."

A Fisher's Solitude

Henry

The second time I saw the fisher it appeared in my headlights on a winter night. Perhaps startled by my truck, it dropped from a tree that leans over our dirt road. Landing in a crouch, it looked right at me. Thick gray-brown fur surrounded its face. Its eyes gleamed white. Sharp teeth sent a warning. A few seconds and it was gone, disappearing over the steep embankment.

I had become accustomed to its evanescent presence, but an actual sighting was rare. I'd sensed it wandering along the tree line near the house and imagined it watching for a careless squirrel, an errant porcupine, or a reckless rabbit. For a while, I thought the dogs' twilight barking was directed only at the squirrels and deer, but I came to believe that sometimes it was aimed at the fisher.

One night, when the dogs were inside, it stole one of the duckness—Disabled Duck—easy prey because he was hobbled by an injury. And so, to me, the fisher became a thief.

But even thieves must eat. And I have little reason to be moralistic here. Plenty of Appalachian families hunted this fisher's ancestors almost to extinction. Their coats were prize fur for the nation's early fashion industry. And as forest habitats disappeared because of logging, the number of fishers decreased. Only in the last decade or so have they returned to the mountains around us. As I contemplated the fisher's plight, a question came to mind: Is survival a matter of luck, and not only a matter of the fittest?

Last fall, Priscilla and one of our sons were riding our horses on the mountain behind us; our two dogs, Viola and Lola, trotted along, as they always did. Suddenly, Viola stopped at a small tree, looked up, and barked a few times. A small fisher clung to a branch ten feet up. Priscilla and our son watched for a minute and then rode on, telling the dogs to come along. But only one did. Viola stayed at the tree: immobile, silent, watching.

And then the fisher made a mistake. He jumped down. Was he too young to understand? Was the danger unseen? Was it just a matter of bad luck? Viola pounced, her paws pinning the little body to the ground. Lola sprinted back and grabbed the furry neck. One violent shake. Then another. And then a lifeless body rested on the leaves. The dogs stayed near for a moment, sniffing the body, and then trotted toward the horses. It was all over in a few seconds.

We were sorry about the death, especially because the fisher family deserves considerable thanks for taking on porcupines. If not for the fisher, my farm might be a more dangerous place for our dogs. Late one evening some time ago, three of them came rushing from the nearby run bed. They were a mess—quills in noses, jowls, and mouths—and whimpering like babes with raw fannies enflamed by seriously soured diapers. They wanted help but didn't want us to touch them. The next morning, the vet pulled out 132 quills across all three patients. Even with the vet's three-dog discount, the bill came to more than two dollars a quill. The dogs laid low for a week, hurting and humbled.

I imagine that if we could tell the fisher our dogs-meet-porcupine story, it would shrug its shoulders. "Idiot dogs," it might say. "Serves them right."

Thinking about the fisher, I ask myself another question: How much creativity does it take to live a life?

The fisher doesn't kill a porcupine as it would a squirrel. Instead, the fisher approaches the prickly job with patience, as a true creative: staying focused, snapping at the porcupine's face, wearing it down until it can't defend itself. As one naturalist describes the process, the fisher finally grabs the porcupine's snout, turns it over to expose the soft belly, and finishes it off. Not exactly neat or quick. But, hey, it's something that coyotes and bobcats won't even try.

Many folks around the farm call the fisher a "fisher cat," but it isn't a cat, and it doesn't fish. Some humans think they hear fishers screech loudly, but it's the fox they hear. Fishers rarely speak above a low squawk. Why do humans get it so wrong? Is it because we need to label mystery? Are we so uncomfortable with the unknown that we grasp for something—a name, a concept, a belief—to help us deal with the ambiguous? And then the story sticks, whether true or not.

The first time I saw the fisher was in the clear-cut above our pastures. I'd gone to feed the horses, telling Priscilla and a few friends who'd come for dinner that I'd be right back. As I closed the barn door, I noticed the fisher scrambling over fallen logs to find safety in the tree line. I followed quickly enough to watch it climb to a crotch of a tall oak. There it sat, looking down. I stood nearby, looking up.

The fisher is a solitary creature, alone for most of its life. The male hangs out with a mate for a few weeks each year but then leaves and plays no role in caretaking. The female feeds and protects her children for a few months and then abandons them to wander the forest on their own—objects of prey for eagles, hawks, coyotes, and dogs. If a young fisher grows to maturity, it faces danger only from humans.

As I watched the fisher in the tree, it remained motionless. It seemed content, unhurried, waiting for no one. Whatever trauma it had witnessed in its life made no difference. Aside from feeding itself and avoiding danger, it had no responsibilities. I was a bit envious of the fisher, and not just because he had the better view.

I started to fidget. I had to get back to the house. My family was waiting for me, but I couldn't bring myself to leave this solitary creature. I was offering it some company, even though it needed none.

The longer I stood watching the fisher, the more restless I became. I'd been gone too long from the house where Priscilla and our dinner guests waited. I appreciate solitude when I'm trying to write, but I'm glad the social world demands my attention. Family and friends have saved me from the depression of loneliness many times.

"So how does the fisher experience solitude?" I wondered as I walked back to the house. It's a question that only a human would ask, and it's not answerable—at least not until we learn to speak its language. I suspect the fisher lives without the burden of consciousness. It is one with its world. When it's frightened, it runs for safety; when it lusts, it finds a mate; when it's hungry, it hunts. It lives alone for most of the year, but does it struggle with loneliness? Probably not. I'm a bit envious.

Back at the house, despite a convivial dinner shared with friends, I couldn't shake a lingering sense of separateness. The fisher had reminded me of the many years I'd spent as a young man feeling lonely. Like many self-reflective, self-absorbed adolescents and young adults, I realized that I was alone in my thoughts: no one would ever truly know what I was thinking and feeling, or so I believed. I embraced separateness—a close cousin to solitude—as an existential fact and resolved to be as emotionally self-sufficient as possible.

Considering myself resourceful enough to survive anything alone was comforting, if illusory. It's painful to remember now, but my favorite song during those years was "I Am a Rock," by Simon and Garfunkle.

Two graces kept me from going down the path to serious depression: a few friends—hanging out meant hanging on—and a lot of reading (novels and science fiction, mostly). Books convinced me that there was a world of experiences, emotions, and situations that might be interesting to explore. Take Salinger's *A Catcher in the Rye*. What struck me as extraordinary wasn't only that a cynical Holden Caulfield could surprise himself at the story's end, but also that some offbeat reclusive writer could create such a complex and relatable character. One person working alone created a community of readers who felt understood. Even then, in some vague and inarticulate way, I realized that writing could both justify and solve the problem of solitude.

In retrospect, it's no surprise that I attach myself so securely to places. They were more dependable than my family in those early years, most of whom seemed to be busy with the need to make a living, taken up entirely with their own emotional struggles, focused primarily on food and gifts as a substitute for affection, or dealing with some relative's prolonged death. Even as a child, I found places to be alone: a coat closet, a stuffed chair on the top floor, a spot under the backyard ailanthus tree. Those places became safe harbors and portals to imagined worlds.

Although unforced solitude has provided me shelter for writing, it has had its dangers. Sometimes, when our marriage demanded more love than I could give, I sought refuge in that old familiar separateness, retreating to that "rock that feels no pain," in the words of my old favorite tune. Priscilla used to call me "cellophane man" when I made sure that all my feelings would slip away and never affect me much. But I was lucky. Priscilla usually found a way to haul me down from that rock to deal with whatever conflict I was avoiding, asking me with her actions whether I could find a way to be truly present even when neither of us were particularly lovable. These days, I'm better at telling her what I'm feeling, even if it's difficult for us. She doesn't use the cellophane tag anymore. Well . . . not often.

Later that night, after our friends left and Priscilla was asleep and I was alone, I realized that the fisher had reminded me of all these tensions in my life: the need to be alone, the fear of being alone, the need for intimacy, the fear of losing myself completely in loving another. How can an innocent creature from the natural world be so provocative?

Last winter I followed a fisher's tracks in the snow to a pile of rocks near a frozen stream. It must have climbed onto a nearby tree because I saw no further signs of its travels. Maybe it was watching me from the crook of a tree. Or maybe it was with its mate. Either way, I'm sure it wasn't writing, and it probably wasn't lonely.

Lovely October

Priscilla

An ache in my back reminds me I've been sitting in one position too long. This bench, here at the pond, is the most uncomfortable seat on the farm. Eddy made it for me years ago. He was proud of his gift, and I never mentioned to him that the bench had issues—like rocking downward without warning, leaving everyone unnerved and grabbing for balance. Eddy and I hauled it out to the pond a few months back and then had to pull it uphill a little and place it at an angle, a little cattywampus. We made sure it didn't put a person in too awkward a sitting position. With a few tries, we finally leveled it out, more or less.

I told him the bench was fine. "Let's just leave it here," I said.

It wasn't any more comfortable, but it was stable. The bench is longer than most, but it's stout and safe for the grandchildren to frolic on it, just as the four-legged kids once did.

I'm sitting on the bench, adjusting my weight, as our dog Lola paddles around the pond. Today is void of human chatter. It's just me and Lola. I like the quiet.

I look up at the hill behind me. The big pines have grown into a thicket. I raise my head and suck in their earthy scent. For my grandchildren, the pines have become a playground, a fort, and a much-needed hideaway during difficult times. They scramble over the "dinosaur skeletons," the needle-less, bone-like branches draping from the carcasses of the huge pines, downed by bad storms years ago. They climb until they are high enough to swing, throwing their feet up as far as they can. Their antics scare Lola to death! She circles underneath them, her head pointing straight up. The children yell as they sail through the air and try to decide where to land. The frantic dog whines and leaps around the acrobats as they hurl themselves skyward.

The trees bordering the pastures have turned into their fall brilliance. Bright yellow maples glow between mahogany oaks and liquid green pines. Leaves loosened from their branches drift through the air, decorating the pasture and floating on the pond.

On this lovely October day, the fields are changing from bright greens to shades of gold and tan as the grasses die back for the winter. The only dark green is the lespedeza that dots the fields. Last year, the goats and cows ate it, but they're gone now so it's come back fast. It's my problem now.

October was always a busy time for me. I was working with our veterinarian to check the stock, and the farmers who had purchased livestock were arriving to pick up their animals. The goats we were taking to the butcher had to be weighed, vetted, and loaded on the truck. I had to complete all the bookkeeping for the year, and I had to make important decisions about breeding for the new year. The ever-present money issues dictated what farm equipment I could repair, buy, or borrow. Forget wants, farming is about needs.

But without the animals, this October is different. The soft wind has a cool undertone that signals a change in seasons. Most field birds have moved on with their new fledglings by now. I'm proud because this year meadowlarks used the untouched fields for their homes. The red-winged blackbirds made their territories in the bushes that surround one side of the pond. Swallows took their nightly part in the wild bird rituals.

I've changed my priorities to include making sure the fields are homes for the wildlife—especially birds. Why not let them rewild the pastures? The pond is alive with dragonflies of all types, frogs, salamanders, snakes, and some native fish, not just goldfish from the pet store. I see more egrets and occasionally a little blue heron or two. This past summer, in the mud along the pond's bank, the grandchildren found more critter tracks of different kinds. One day they saw four large paw prints, two deep in the mud and two farther up the bank. It must have been a bear! We never had proof, but we were sure it was a bear.

I adjust my position on the bench to relocate the pain in my back and comfort myself by staring at the maple tree on the other side of the pond. It's brilliant with its crown of reds, oranges, dusty browns, and golds. But the lower half of the tree is still a fresh summer green. The pond's surface holds the tree's reflection perfectly. It is beautiful.

The colors in these eastern mountains, especially in October, are always inspirational. Henry and I, with our sons and a dog or two, walked the mountains of West Virginia for years before we bought the farm. I took hundreds of photos of the mountains' unique colors and textures; the photos would hang on the walls of my studio in Baltimore. Those colors never failed to make their way into new collections of clothing and scarves that I designed and painted as I prepared to do battle for yet another season in the arena of New York City's fashion world. It was another world, another time—but why does it feel like it was just yesterday? Time doesn't march on, it races.

My back aches. I've sat too long on this uncomfortable bench, but I'm not ready to leave yet. I look at the maple tree and the still pond.

Alone at the pond I don't feel lonely. I feel the spirits of our families sitting here and walking the fields with me. This place, this keep, was born because of a family's commitment to family. It's a place of safety I can always count on.

The air sharpens into the damp cold of an October evening. The light fades. Shadows grow. The sun is sinking into pink and red glory behind the mountain. The reflection of the maple tree is just about gone from the pond.

"It's near dark, isn't it, Lola?" I whisper as she sits patiently, waiting to meander home.

Finally, I rise from the bench, holding my aching back. Lola and I make our way toward the house, content in knowing we are making our way through the fields of the keep.

Epilogues

The Near Final

Henry

As a skinny ten-year-old, I'd bother family members with random questions about life. Where did we come from? How many stars are there, really? What does it mean to be kind? Most adults treated my curiosity as an irritating interruption to whatever they were doing.

One uncle was different. He listened to me and sometimes offered a few answers. He was the only family member who understood that I couldn't help myself. I wanted to know things, especially the whys.

The quest to resolve persistent mysteries about our existence is to search with slim chance of success. But still I have faith—not so much that I'll find an answer, but that the searching itself will be redemptive. It's always been so for me. Long ago, I accepted mystery as an abiding part of my life. Mystery as my story.

During the summers of my late adolescence, I was lucky to canoe the rivers and lakes of northern Canada with a small group of teens and a few adult guides. Four summers in a wilderness that no longer exists gave me a new appreciation for mosquito venom as well as enduring memories of spumy haystacks in roaring mile-long rapids, the weird beauty of the northern lights, and the howls of a coyote under a full moon. During these trips, I filled journals with earnest descriptions of transformative experiences and asked myself all the questions I could dream of. I never found another place that so provoked me to write until we found our farm.

As soon as I set foot on our land, I understood that it contained all sorts of mysteries; experience slowly transformed many of them into something else.

Some mysteries metamorphosed into knowledge. The puzzle of the tractor's ruptured pressure line led to a deeper understanding of hydraulics; the challenge of screwing rectangular sheets of rooftop tin to the parallelogram of the barn's frame forced me to acknowledge basic geometry; and the futile attempts to catch a two-month-old goat by hand yielded, eventually, to the wisdom of using a shepherd's staff. It's called learning, and I'm glad for the opportunity to learn stuff I never thought I'd need to know.

And then there were mysteries that experience translated into faith. How exactly does a mundane sunset turn into something beautiful, and how does the evening light make the beautiful breathtaking? No matter how many times I experience those sunsets, I don't have an answer, just faith that some answers will emerge from the human imagination.

Here on our land, I've lived much of my time in the space these kinds of questions create—the aisle between the mundane and the mysterious. Living in this liminal space has required finding faith in a story larger than my own. And that larger, longer story involves the power of storytelling itself. Born in the human imagination, storytelling has birthed in me a faith that the world maybe makes sense after all: at its best, it offers clues, and occasional answers, to stubborn mysteries.

Listening to stories in a group—and especially telling stories to a group—has particular power. The strength of twelve-step programs resides in speaking one's own story, sometimes over and over, to a group of skeptical sufferers, while accepting the grace offered by those who hear your story and understand it as a part of their own. The performative aspect of storytelling—whether it's playing in a Shakespearean tragedy or speaking slam poetry or reading an essay-in-progress to fellow writers—brings the written word to light and to life. I love reading my stories about this place out loud almost as much as I love the place itself.

When Priscilla and I began reading our stories to each other, we found we were exploring in words the land's pathways and cul-de-sacs that we had previously traveled together on foot. We learned what we didn't know the other had seen or experienced, and those insights were sometimes unsettling.

When she read her first draft of the fire story, I finally understood that her anger at me was connected to her fear of being alone, of having no one there for support when things went bad. That fear is deeply rooted in her experiences as a child, when neither of her parents provided the reassurance and acceptance that every child needs. I should have seen this connection long before, but my blinders curbed my vision. Perhaps I didn't want to acknowledge her neediness for reasons that have to do with my own upbringing. For both of us, childhood trauma has long legs.

It went the other way, too. Priscilla's reading of "Lucy" renewed my deep respect for her extraordinary intuition.

Because I'm hobbled by a mind too caught up in meaning-making, I treasure her ability simply to watch carefully and understand an animal's experience of the world. Where I would see two goats grazing next to each other, she'd see an older mother comforting a younger doe who'd just lost her baby. Many of her stories reminded me of the mystery of intuition.

When I read my stories to her, I'd brace myself for her reaction.

"Where's the blood?" she'd sometimes ask.

"What do you mean 'blood?'"

"You've always told me that good writing means you gotta spill some blood on the page. I don't see any blood. It's just blah-blah-blah."

"So, okay, I need to put some more emotion into the story. Can you point out where?"

"You've got to figure that one out. It's your story, not mine."

It was never easy. I learned always to label the file of the story I read to her as "near final" because her reactions inevitably led to changes in my writing—and often in my understanding of myself. Perhaps every version of every story we tell, just like every version of ourselves, is a near final.

In the years ahead, I may stumble on solutions to a few mysteries that puzzle me, but there's one question I will never answer: How and why did it come to pass that a Brooklyn boy and a Georgia girl discovered each other long ago and later found a farm near the edge of the wild and still later wrote some stories and still later assembled them into a book? I can relate a fact-filled story of the how, but the why will remain unknown forever. That our love for each other has endured for so long strikes me as one of the universe's greatest mysteries.

Letter to My Mother-in-Law

Priscilla

Early morning mist
Doesn't dampen their cries as the truck rolls off.
The last one takes
Does and best dog to new lives,
Leaving pastures to the quiet sadness of those left behind.
Mist closes around me. It will blow off soon.
I need to get on with this.

Dear Alice,
I know you had a lot of sadness in your life, so you'll under-
stand. I got to the day when I couldn't do it anymore.
The turning point was a surprise three-foot snowfall in early
March, followed by a week of bitter cold. I had to dig out gates and
make paths for the goats with a shovel and the tractor. I worked for
hours in frigid weather, tending livestock, making sure they had
water and extra hay, and creating paths to walk out of the barns. I
was beat and could hardly climb down from the tractor at the end
of that day. Every joint in my body screamed in pain. I thought
the arthritis had defeated me many times before, but somehow, I
had kept going.

This time was different. I couldn't imagine how to continue. I came into the house with tears streaming down my face, trying to peel frozen gloves from hands that wouldn't move. I didn't feel my feet and I knew thawing my body was going to be painful. I sat in front of the woodstove, slowly removed frozen clothes, cried a bit, and drank whiskey. "What do I do now? Will I hurt this bad in the morning? Shit!"

After a shower, I drank hot tea, took three Advil, and sat, still miserable from pain and exhaustion and shock. My doctor had told me the arthritis was worsening. As if that was a surprise to me. I was losing against my body, again, and I felt sorry for myself.

I know, Alice, you'd say "I'm so sorry. Do you want to keep going? What about your family? What about the animals? What about you and Henry?"

As my hands and feet warmed, they began to ache. I had to just deal with this and wait for the ache to go away. It did—after another shot or two of whiskey. I reminded myself that I'd accomplished what I needed to do for my animals that day. They were safe, dry, and well fed, with water that wouldn't freeze too quickly. I could rest now; everyone else was okay.

The next night your son came home from the city. We had a serious talk, and after hours of discussion, we agreed it was senseless to go on farming. I was in over my head, and we couldn't find the right workers to help me. Eddy was as old as I was, and his tough life had left health hardships on him also. He'd come and help if I called him, but I didn't want to burden him with constant work.

The horrible sick feeling of loss was settling into my gut. I knew the feeling well and I was crushed to realize I was going through it again.

I was staring decrepitude in the face. Just as when I gave up my designing, work I loved—something that made me who I am, a good farmer and breeder, complete—had become impossible for me. Although it'd been exhausting, the work had an exhilarating, easy rhythm. The commitment I made to the livestock—from goat to cow to duck—was joyful. If it weren't for the damn disease that plagues me, I'd still be pitching hay.

Would you have preached keeping on? Would you have said, "You won't be who you are if you give it up?" Maybe, and maybe your words would have made a difference, but you're not here and all I can do is imagine them.

When times were tough, I took the no-nonsense advice you had often given me: "Remember why you do your work and get on with it." I walked my pastures twice every day, notebook and sharpies in my coat pockets, and at least once every night before I went to bed. The medical bag held tags and tag gun, a scale, wound dust and purple antiseptic spray, gauze, duct tape, and a flashlight. I wore waterproof boots and carried a big stick to defend myself from those crazed first-time moms. I looked for new babies and birthing does who were struggling.

I stalked the fields until every kid had been born, weighed, tagged, and listed in my herd book. That time of year was the best for me. It was why I did the heartbreaking, mind-numbing work of farming. Most of the time we were doing a good job and thank heavens we didn't need to make money off the farming. But my body was in pain. Always.

In the spring we made final plans for closing the farm. I never knew if Henry was relieved or angry. It's like a death when a farmer stops farming. I wasn't comfortable announcing my decision, but

word of the farm's closure made its way slowly through my network of farmers and breeders.

I had amassed a good herd of heritage Spanish meat goats
and Dexter cattle. Henry and I wanted to make sure we found
the right place for all of them—programs and farms that would
respect the herd's genetics. Over those summer and autumn months,
the farmers came, and I was able to usher the cows and goats into
parts of the heritage world that would carry on the work of con-
serving and promoting breed diversity.

The fields are empty of livestock now. They are too quiet. The
animal sounds are from wildlife and a couple of horses. No goats
or cows or hogs. The drama's gone. The only drama was from Henry's look as he surveyed the empty pastures, shoulders slumped.

Alice, I watched you keep your world together, mainly for the next
generation. But also, we both knew that the other side of it was
making sure you could stay in your keep, to continue to live in the
old house around objects that had been in your family for generations and had surrounded and comforted you for your whole life.
The animals that are now gone had been my comfort. I wish my
parents had saved more objects from their families—paintings, oil
lamps, tables, whatever. You kept things and wanted us to treasure
them as objects with special histories of their own. And we have.
We have your old sideboard in our living room, and your dining
room table around which we gather every day. These things are a
bit worn and cracked, but they're still collecting history. My animals were not worn and cracked; I was. Anyway, I never asked
you about what it meant to have all those things. Would we have
been able to talk honestly about that?

When we decided to stop the farm operations, we had three livestock guardian dogs: Arnost, Bella, and Sophie. Arnost was most upset about the departure of the farm animals. He was not happy without his job. As much as we would miss him, we gave Arnost to a couple who needed a guardian for their herd of forty sheep. They were kind people and had been losing lambs to the coyotes. Arnost would have the life he loved the most—taking care of his own herd. But it was tough to see him leave. He didn't want to get into the back of their van. And, when he finally did, he looked back out the window. His sad eyes spoke clearly: *I don't understand. This is my home. I had a job. I did it well. Why are you sending me away?*

Sophie and Bella were Maremmas who had worked together for years. They stayed with us on the farm after the goats left. One slept on the hill behind the house and the other by the dirt road running past our driveway. Sometimes in the mornings, we found the two dogs in the pasture with the horses, protecting what was left to protect.

The dogs were ecstatic when your great grandchildren came to visit; they had their little, noisy, two-legged humans to keep watch over. We always knew where the great-grandkids were because the dogs were there too. Your city-dwelling granddaughter loved to tell her children to go play, adding "And make sure the dogs are with you." Of course, none of your great grandchildren will remember the livestock and the dogs in a few years.

And now, just Lola is left. She tries to give them her all.

In the early spring after the last goat was gone, I watched the long grass as it waved in a warm wind. A couple of meadowlarks flew out of the grass. Now, I focus on what the land can do for its native *inhabitants. The wildlife that's left can reclaim their homes in the fields when they need them most. The farm buildings can become their keep through the harsh West Virginia winters. I found holes dug under the goat barn. Paw prints showed that many of our wild neighbors were coming back.*

Last year, two deer and their fawns lived along the tree line behind the house. I'd see them crossing the big field early in the morning or late in the afternoon. The two fawns played and bucked in front of their moms. Most afternoons, the small parade made its way to the old pear tree on the hilltop. The big field was an easy place to watch for predators so they stayed there for a while, eating every pear they could reach and enjoying the rich grass that was all theirs.

When I raised goats, the deer were enemies that carried diseases. Now I watch the fawns frolic and think of the field as a pasture for wildlife.

Maintaining fields that are not in use is a tiring job. The goats fought back the briers and other weeds for years and the cows kept the grasses pounded down and reseeded, and both helped fertilize the fields. Now only two horses roam the large pastures, so the multiflora rose and the broom grass have come back with a vengeance.

When a farmer up the hollow asked if he could take the hay off the fields, we agreed. We'd share the cost of lime and fertilizer. The farmer would cut the fields the following week. I always made him wait until late July because I wanted the field birds to finish sitting their nests and to start flying with their new broods. The farmer wasn't happy about cutting so late in the hay season, but these are my fields—and my agenda goes beyond just cutting hay. Besides, he was getting the hay for free.

That is, a farmer's "free." He still had to work a solid week spending at least ten hours a day in the hot July sun, putting up hay with his temperamental, ancient machinery and worrying about the threat of bad weather. That's the kind of "free" hay the small farmers have around here. But at least he could feed his small herd of cattle in the winter. He loved those cattle, even though he couldn't really afford them and was too hobbled by age, bad health, and lack of money to care for them well. But he needed his dreams. I understand.

Age and bad health eventually caught up with him, too. He sold all his cows. He's still getting around, but he's lost some of the sparkle in his eye. I understand that, too.

One crazy thing is that I was well-suited to be a farmer. I never minded the work and was able to move through the depression that came with death, mistakes, disappointments, cold feet, and frozen hands. Those things didn't matter because the next morning, I'd be out with the herds and some kid or calf would look over at me with a plan for morning mischief, and all I could do was think of myself as a very lucky woman, working in the most beautiful office in the world. I loved this farm and farming. Viscerally. The hardships couldn't sway me from the commitment it demanded. Until they did.

Alice, I'm sorry to be rambling. But I miss it so.

Henry reminded me that I was able to ride our horses more often. For me, that is a big plus. One autumn day not long after all the stock were gone, I swayed in the saddle to the slow rhythmic movements of my horse, Wid, as we walked up a mountain trail. We had brought our horses to ride in the Virginia Highlands, one of our favorite places to horse camp. The trip had given us some time away from the empty fields and quiet barns. With weepy eyes, I tried to imagine myself with a new job, a new identity, a new purpose. A different me?

I looked at the sky. The sun trickled through openings in the canopy. Leaves floated to the ground, creating a broken carpet of brilliant yellow and red patterns as they lay on the dark tan of the dirt trail. The shocking blue sky exposed a vivid backdrop for the hot colors of the leaves that still clung to their glory. Wid's hooves thumped in a four-beat. The day could not have been more beautiful, I was with my man, and I was happy.

My big gray horse helped heal my wounded soul by carrying me up many mountains as I made my way on my mental journey, asking nothing of me but an occasional carrot. One of my favorite sayings of Teddy Roosevelt's was, "There is nothing better for a person's soul than the back of a horse." Wid must have sensed my depression and fragility, or maybe my training had finally paid off, because he'd became gentler and more attentive to my commands. We taught each other patience. The solitude of our rides gave me the space that helped me understand that I had entered a new chapter.

Saying goodbye to my farming took a hunk out of me and hurt as much as giving up my designing. But I'm not dead, so I need a plan, if not just for sanity.

"Oh, get over it and move on," you'd say with a certain finality. "Just turn that corner."

I can see you staring right into me. We were close, weren't we?

"Stop with the weepy eyes. You had a good run," you'd insist.

Your son once said something similar, with a kind smile: "You can still ride a horse for two and a half hours without getting off. And you have more time to spend with our grandchildren."

So, you and he are right. It's another chapter. I'll explore new artistic challenges. Take a walk just because I want to take a walk. I didn't do that when I farmed. There was never free time; farmers don't have free time. If I wasn't doing, I was thinking about doing.

I haven't answered the question that I still ask myself: What is a farm when it is no longer a farm? I still walk these fields with a deep need for our little bit of the mountains and our high meadows. I know every turn on our land, but it never ceases to amaze me how new it looks through the years. I can't imagine calling anywhere else "home," "the farm," or "my place."

The farm is our keep, and I'm safe to explore.

Goodnight, Alice.

A deeply grateful Priscilla

Acknowledgments

....

For better or worse, we dragged many of our friends into the process of writing this book. Some read early drafts of particular essays, and a few kind souls provided comments on large portions of the text. Some friends were close by as the stories happened. Some listened with a certain bemusement as we told the more tortured story about the writing process itself (told twice over, in different voices). A deep hug of gratitude to everyone who supported us along the way.

Particular thanks to Winfield Swanson and Martha Bustin for each editing the whole text and for their ongoing support and friendship; to Cathie Gandel for wise guidance at key turning points; to Sandy Tritt for early and steady encouragement, and for giving Priscilla the safety to scream on a page and then delete it and go on; to Mac Hart for the honesty, humor, and love that comes only from a deep and steadfast friendship; to Betty Puffinberger for listening to Priscilla's stories, remaining patient through all the blah-blah-blah, and editing kindly; to Joel Bell for an authentic enthusiasm that knows no bounds when it comes to friends and to his wife, Ellen, whose quiet support helped immensely; to Gabe Gopen for astute comments that encouraged Henry to keep going; to Deborah Fugate for both her unalloyed ebullience and sharp editorial eye; to Jaimie and Mimsy Stirn for recognizing that Henry has always wanted to do something like this; to Justine Payton for helping us believe our stories might actually have some depth; to Su Liu for wanting to read the first few essays and then wanting to read the rest.

Joe Razes is a wonderful friend and his honest pictures of the farm were a deep well for our imagination. Many thanks to Lisa Elmaleh for capturing the essence of us and our place through her photographs.

We are deeply grateful for the support and guidance provided by the staff at the West Virginia University Press who invited us to work with them and helped bring the book to publication: Sarah Munroe, Than Saffel, Marguerite Avery, and Kristen Bettcher. It's been fun!